Uses of Energy Minerals and Changing Techniques

Uses of Energy Minerals and Changing Techniques

Kaulir Kisor Chatterjee
Chief Mineral Economist (Retd.)
Indian Bureau of Mines, Nagpur-440013

PUBLISHING FOR ONE WORLD

NEW AGE INTERNATIONAL (P) LIMITED, PUBLISHERS
New Delhi • Bangalore • Chennai • Cochin • Guwahati • Hyderabad
Jalandhar • Kolkata • Lucknow • Mumbai • Ranchi
Visit us at **www.newagepublishers.com**

Published by New Age International (P) Ltd., Publishers
First Edition: 2006
Reprint: 2009

Branches:

- 36, Malikarjuna Temple Street, Opp. ICWA, Basavanagudi, **Bangalore**. ✆ (080) 26677815
- 26, Damodaran Street, T. Nagar, **Chennai**. ✆ (044) 24353401
- CE–39/1016, Carrier Station Road, Near Paulson Park Hotel, Ernakulam South, Cochin ✆ (0484) 4051303
- Hemsen Complex, Mohd. Shah Road, Paltan Bazar, Near Starline Hotel, **Guwahati**. ✆ (0361) 2543669
- No. 105, 1st Floor, Madhiray Kaveri Tower, 3-2-19, Azam Jahi Road, Nimboliadda, **Hyderabad**. ✆ (040) 24652456
- RDB Chambers (Formerly Lotus Cinema) 106A, Ist Floor, S.N. Banerjee Road, **Kolkata**. ✆ (033) 22275247
- 16-A, Jopling Road, **Lucknow**. ✆ (0522) 2209578
- 142C, Victor House, Ground Floor, N.M. Joshi Marg, Lower Parel, **Mumbai**. ✆ (022) 24927869
- 22, Golden House, Daryaganj, **New Delhi**. ✆ (011) 23262370, 23262368

ISBN(10) : 81-224-1803-1

ISBN(13): 978-81-224-1803-3

Rs. 195.00

C-09-08-3925

Printed in India at Glorious Printers, Delhi.
Typeset at Kalyani Computer Services, Delhi.

PUBLISHING FOR ONE WORLD
NEW AGE INTERNATIONAL (P) LIMITED, PUBLISHERS
4835/24, Ansari Road, Daryaganj, New Delhi-110002
Visit us at **www.newagepublishers.com**

"Strange how much you have got to know,
before you know how little you know."

Dedicated to my inner God

PREFACE

When I was a student of Applied Geology four decades ago, while I could find literature listing the various economic uses of minerals, my thirst for learning the reasons why only a particular grade of a particular mineral should be suitable for a particular use, and not any other mineral, or what happens if some other grade of the same mineral is used, what were its substitutes, were there ways and means to ensure total usage without rejecting anything, etc. remained insatiable. Nor my teachers could provide much guidance in these regards. Cramming by rote the lists without understanding the fundamentals, somehow saw me through those years in the university. Few years later, when I was working in the Mineral Economics Division of the Indian Bureau of Mines, I began getting more and more exposure to the economic side of minerals. It is then i.e., about three decades ago, that I first got the idea of treating the subject of uses of minerals from a different angle, and I started going into depths of the subject and collecting information from all possible sources—not only the available literature on different related subjects, but also directly through interactions with industries and institutes. After 1991, when Tim Berner-Lee invented "world wide web", the same also proved to be of great help to me in giving concrete shape to my idea. I decided to address the most important group of minerals from the point of view of economic significance, namely the 'energy minerals'.

I have endeavoured to move away from the more conventional method of treatment of the subject and in my endeavour, many persons have inspired and helped me in many ways. I am thankful to them. Through this book, I hope to take the readers to the depths of science and technology of utilization of the energy minerals, without making it a highly specialized one in any way. But, the readers may bear in mind that the uses of minerals, as the title indicates, are with reference to *changing* technologies; and change is a continuous endless process. This book takes into account the status up to the beginning of 2005.

— KAULIR KISOR CHATTERJEE

CONTENTS

	Preface	*(vii)*
1.	**Introduction**	**1 - 4**
2.	**Coal**	**5 - 51**
	— History	5
	— Criteria of Use	7
	— Uses	12
	— Specifications of Use	13
	— Utilization of Wastes	24
	— Substitution	38
3.	**Petroleum**	**52 - 74**
	— History	52
	— Criteria of Use	54
	— Common Uses	57
	— Specifications of Use	62
	— Utilization of Wastes	70
	— Substitution	72
4.	**Natural Gas**	**75 - 83**
	— History	75
	— Criteria of Use	76
	— Common Uses	77
	— Specifications of Use	81
	— Utilization of Wastes	82
	— Substitution	82
5.	**Uranium Ore and Metal**	**84 - 96**
	— History	85
	— Criteria of Use	86

	— Common Uses of Ore and Metal	89
	— Utilization of Wastes	94
	— Substitution	95
6.	**Lignite**	**97 – 99**
	— Criteria of Use	97
	— Common Uses	97
7.	**Monazite and Thorium**	**100 – 108**
	— History	100
	— Criteria of Use	101
	— Common Uses	105
8.	**Peat**	**109 – 110**
	— Criteria of Use	109
	— Common Uses	109
	— Substitution	110
9.	**Anthracite**	**111**
	— Criteria of Use	111
	— Common Uses	111
10.	**Bibliography**	**112 – 113**
	Glossary	**114 – 115**
	Index	**116 – 123**

CHAPTER 1

INTRODUCTION

Use of a commodity depends on the interplay of its demand, supply and technology; and the use manifests itself through the consumption of the commodity. Commodities may be defined in different ways at different times, new properties may be unravelled, but the uses of the commodities follow their own evolutionary course depending on man's knowledge of their properties at a particular point of time. It is the potential utility, which has given rise to the demand, and has acted as the motive force behind all economic, scientific and technological activities.

Use, manifested through demand and consumption of a commodity, and substitution go together; and are sometimes a continuous process while at other times a cyclic process. Yesterday, we might have used firewood as a source of energy, today, we may be using coal or oil, and tomorrow we may switch over to nuclear energy, solar energy or some other non-conventional energy. But, who knows, one day one of us may wander into a situation where nothing but firewood is available, and one may be forced to use it again! It all depends on availability and economics—particularly when the commodity is a natural one like mineral, and not a man-made one.

However, availability or quantity of a commodity is not the only dimension in its use and substitution. There is another dimension, viz. quality. In fact, quality and quantity have a sort of complementary relationship inasmuch as improvement in quality usually tends to reduce the quantity of consumption and vice versa. But one important difference must not be overlooked. While quantities can be measured in terms of universally accepted units like tonnes, kilograms, liters, barrels and so on, there is no such standard for measuring the qualities of commodities. This problem of evolving some universally acceptable standard for evaluating qualities is perhaps the most apparent (and also widely debated) in case of mineral commodities, because of mainly two reasons. Firstly, the minerals are the basic raw material for practically all goods—either directly or indirectly. Secondly, the minerals being endowed by nature without any control whatsoever of man, are extremely variable and unpredictable in quality.

It is obvious that there has to be a set of standards not only for each commodity, but (in case of raw materials in particular) also for each end-use. The numerical values of different parameters of quality usually signify the trade-off amongst the different conflicting interests and perceptions of buyers and sellers; and in individual situations, local adjustments and compromises depending on exigencies of market are always possible as well as practicable. In other words, there may be some subjective consideration in deciding the numerical values of various desirable and deleterious parameters in a commodity, which may vary from time to time and from plant to plant. But the reasons underlying desirability or otherwise of any parameter are dictated purely by technology.

Mineral raw materials are used in mineral-based and mineral-consuming plants for making various products, the properties and functions of which are predefined. Primarily, it is the physical properties and the chemical composition of a mineral commodity that render it useful for obtaining a particular product. But apart from the intrinsic nature of the mineral commodities, there are some external factors influencing their usage.

Firstly, the required mineral commodity must be available to the consuming plant at an economic cost. Fertilizers can be produced based on either coal or natural gas. If natural gas is not available economically to the fertilizer plant, coal may be used and vice versa. Secondly, the interrelated raw materials should also be available in the right grade and at the right cost. For any product, invariably a host of raw materials are used. Unavailability of one may affect the use of others. Thirdly, the intermediate products (in between the finished products and the raw materials) as well as the finished products must withstand the pressure of substitution. Fourthly, the use of mineral must not cause too much environmental pollution. Fifthly, the wastes generated directly or indirectly due to the use of mineral raw material, must not cause irreparable damage to ecology, and disposal problems should not be overriding. Finally, utilization of a mineral is never total, unless the wastes generated at different stages of its use, can also be utilized and recycled back into the economy.

Summing up, it may be said that for a particular product, a particular mineral raw material will be used, when the latter is superior to all its potential substitutes at every stage of the whole material cycle starting from physico-chemical nature to recycling or disposal. On the other hand, if a mineral loses its superiority to another material, it will tend to be substituted. In fact, the use and substitution of any commodity at any given point of time represent an equilibrium between the force and counterforce within an economic system. It is in this context that we have to look into the question of why a particular grade of a particular mineral commodity is used for a particular end-product. An understanding and appreciation of the uses of minerals from this point of view are generally not seen.

Right since the time the first man came on the earth, minerals have been directly or indirectly supporting human life and civilization. The principal natural resources of economic significance are the human intelligence, animal energy, forests, water, land for agricultural activity and minerals. But minerals are required for the nourishment of forests, for fertility of soil and consequent food production for men and animals, and also for satisfaction of different types of human want. Thus, minerals form the backbone of life on the earth. In fact, the minerals were there on this planet when no form of life could start. And as for the present day human society, not a single day can be passed without using minerals and metals in one way or the other. Right from the pin to the spacecraft, from farming to education—everything and every activity that we are concerned with, requires minerals and metals. Take away mineral from life, and its omnipresence will at once reveal itself.

And amongst all the minerals, the energy minerals are receiving more attention of man at present than the others. This is because, for almost any economic activity, and even for using other minerals to transform them into some useful products, energy is needed; and the primary role of energy minerals is to provide that all-important energy. In this book, those energy mineral have been selected for focusing the readers' attention.

To classify minerals into groups based on use is not the perfect way to do it, and such groupings are, more often than not, beset with problems and ambiguities. A mineral may be used for a number of purposes and for making a number of products, but by considering the somewhat major or the most obvious use of each individual mineral, it has become a matter of convenience to regard minerals as 'energy minerals' or 'metallic minerals' or 'industrial minerals'. This basis of grouping is followed in so many mineral statistical reports and publications, that by now it has practically become an established convention. Although it will be pointless to seek to arrive at a standard definition, presently this is an accepted practical classification based on economic considerations.

In Indian context, by energy minerals, traditionally, people used to understand coal and petroleum. However, in the post-independence era, rapid changes in the understanding of people have taken place. Firstly, the discovery of large resources of lignite in Neyveli in Tamil Nadu (where there is neither any coal nor any petroleum resource), and coming up of large lignite-based thermal power plants there, have suddenly transformed this mineral from a useless substance into an economically significant energy mineral.

Similar is the story of natural gas. In India, natural gas is produced as a by-product from crude petroleum wells, and during the greater part of the history of petroleum production, this by-product used to be burnt out. It is only after the oil crisis of 1972 (when petroleum prices suddenly increased many fold) that industry's attention was turned towards the potentiality of natural gas not only as a substitute fuel, but also as a source of industrial chemicals. Now, even pure natural gas fields are being tapped besides utilizing the gas produced as a by-product from oil fields.

Uranium shot into limelight after the first atom bomb was exploded in 1945. But during that time, Kapitza, a Russian nuclear scientist wrote from Siberia to one of his friends: *"To speak about atomic energy in terms of atomic bomb is comparable with speaking about electricity in terms of electric chair"*. As rightly foreseen by him, as well as by some other scientists, atomic or nuclear energy has become an important component of the economies of many developed countries. India entered the nuclear age in early post-independence era when the first Indian atomic reactor was commissioned. Uranium ore was traditionally regarded as the most important, if not the only, economic source of nuclear energy. Unfortunately, there was only one uranium mine in India in Jaduguda, Jharkhand, and even presently this continues to be the situation. Relentless efforts to discover new economically minable uranium ore deposits and open new mines, have not borne any success. But at the same time, there are abundant resources of monazite, a thorium mineral, in the beach sands of India. Although, theoretically, thorium was known as a nuclear energy mineral, until recently, there was no feasible technology available. Now, after the development of the fast breeder technology, thorium is becoming a practically usable nuclear fuel in India.

Strictly speaking, peat and anthracite are not energy minerals from the Indian point of view. However, in some other countries, anthracite is used as a major energy mineral, and peat as a minor one. Moreover, these two minerals constitute the two ends of the coalification process, peat being the first recognizable stage, and anthracite, the last. Basically, these two minerals are carbon, like lignite and coal. Hence, it has been thought logical not to separate them from other energy minerals and group them together instead. Moreover, as is evident from the foregoing paragraphs, the usage of minerals is highly dynamic, being dependent, apart from on the taste and living standard of man and on technological developments, also on the outcome of newer and newer exploration efforts and researches in the fields of mining and mineral beneficiation. If some grade or some mineral is not used today, who knows, it may become an indispensable commodity tomorrow!

In the final reckoning, therefore, all these eight minerals comprising the traditional and the most modern energy minerals have been included in this book. It is again emphasized, however, that although these are being regarded primarily as energy minerals because of their most immediate and obvious usage, like any other mineral, these also find multiple applications; and in this book all the uses are discussed.

Since, as has already been mentioned, use of a mineral cannot be considered in isolation of its economics involving waste utilization, substitution and, of course, the physico-chemical criteria of use, all these aspects have been addressed in case of these minerals as far as relevant. To serve the wider interest of the general readers, history of usage of the minerals has also been traced as far as possible.

It has also been presumed that all the readers may not have very specialized knowledge of the subjects of physics or chemistry. Hence, as far as practicable, the scientific terms pertaining to those subjects have been explained to facilitate understanding by a wide range of readers.

2

CHAPTER

COAL

Geologically, coal is a complex substance derived from buried plant material which underwent alteration due to heat, pressure and chemical and biochemical processes. The character of coal depends on the type of the original plant debris and the amounts and duration of heat, pressure, alteration, etc. The process of transformation of the source material into coal might have been completed or arrested midway, thus giving rise to coals of varying maturity or 'rank'. In order of increasing maturity, coal is ranked as peat, lignite, bituminous coal and anthracite. However, in commercial or economic sense, by coal one normally means the bituminous rank.

In most literature, coal is referred to as a fuel or energy mineral. But we shall see in this chapter that now its use for manufacturing a host of chemicals is no less important than that for producing heat.

HISTORY

Firewood had been serving as the sole fuel for centuries and millenia ever since the day when man learnt how to light a fire. It is surprising to think that coal, which is one of the most common commodities in everyday usage today, was not known to the ancient man who nevertheless knew about so many metals and alloys starting from copper (4000 B.C.), bronze (3000 B.C.), and iron (1800 B.C.). Firewood was the cornerstone in metal-smelting operations, and much later wood charcoal came to be known as a better substitute of wood. In 1450 A.D., when the first blast furnace was set-up in Great Britain, charcoal was the fuel and reducing agent used in it. However, during the following years, the proliferation of iron furnaces in England and the consequent cutting of forests for producing charcoal assumed such a magnitude that the British Parliament had to pass an Act prohibiting further expansion of the industry. But during the period from middle of 16th to late 17th century, fire-dried chopped wood (called 'white coal') was used for lead smelting, because thermal value of charcoal was high enough to evaporate lead.

It was only in 1621 that the iron-smelting industry received an impetus when Dud Dudley discovered 'pit coal' (viz., the coal recovered from pits) as a viable alternative to charcoal. But the art of smelting iron with pit coal was forgotten after the death of Dudley in 1684, and was revived only during the period 1730–35 by Abraham Darbys. This marked the real turning point in the history of use of coal and a great landmark in the history of human civilization. By the middle of the 18th century coal was replaced by coke in iron-smelting. Thereafter, coal started finding application in steam engines and steam ships, thus revolutionizing the transport system. Though a steam-operated machine was built in 1698 by Thomas Savery for pumping water up a mine shaft in England, and a steam boat was built by Thomas Newcomen in early 18th century, the first steam engine was invented by James Watt in 1769, followed in 1784 by a double-action steam engine. In 1803, Robert Foulton tested his first steam ship in Seine river in Paris, while in 1814, George Stephenson built the first steam locomotive. Meanwhile during 1792-96, William Murdoch laid the foundation of the coal gas industry in England. By distilling coal in a closed iron retort, he produced coal gas for use in an indoor illuminant. In 1810, supply of coal gas was started in London on a commercial basis.

During the late 18th and the early 19th centuries, coal and iron together revolutionized the canal and railway transportation system in Great Britain.

In USA, the first coal mine started in 1745 near Richmond, Virginia. It was a bituminous coal mine. First anthracite mine was set-up in 1793 in Lahigh, Penn. Ever since that time, the output has been steadily increasing due to coming into being of more mines. By the end of the 19th century, coal firmly entrenched itself in the industries of several other countries like Germany, France, Belgium, Russia, India, etc.

In India, coal was known to occur in Raniganj area as early as in 1774 and was actually worked for the first time in 1777. However, regular mining of coal did not start before 1814. Commercial coal mining commenced still later—in 1828, when a new company named "Carr & Tagore Coal Co." was set-up (Prince Dwarkanath Tagore, grandfather of Rabindranath Tagore was one of the partners). Prior to 1855, coal used to be transported from Raniganj to Kolkata along Damodar river by boat. In 1855, the Kolkata-Raniganj East Indian Railway was completed and coal production received a boost.

In India, coal replaced charcoal in iron-making for the first time in 1875. In that year, two blast furnaces were built in Kulti in the Raniganj coalfield for producing pig iron with the help of coke.

By the year 1885, there were as many as 68 collieries, and this number swelled to 123 during the subsequent decade. The production of coal in India since 1880 is as follows:

Year	*Coal production in India*
1880	About 1 million tonnes
1890	Over 2 million tonnes
1900	6.22 million tonnes
1910	12.97 million tonnes
1920	18.25 million tonnes
1930	24.18 million tonnes
1940	29.86 million tonnes
1950	32.82 million tonnes
1960	52.59 million tonnes
1970	73.70 million tonnes
1980	109.15 million tonnes
1990	201.82 million tonnes
April, 2000-March, 01	313.70 million tonnes
April, 2002-March, 03	341.25 million tonnes

Note: Production in Pakistan included up to 1940.

CRITERIA OF USE

There are certain physical and chemical characteristics of coal which stand out prominently in its practical use. These are:

1. Chemical composition
2. Thermal value
3. Fuel ratio
4. Reducing property
5. Coking property
6. Weatherability
7. Specific gravity
8. Abrasive power
9. Grindability
10. Ash fusion temperature
11. Gamma ray absorption.

1. Chemical Composition

The chemical analysis of coal shows that it consists principally of carbon, hydrogen, oxygen and earthy matter, and also small quantities of nitrogen, moisture, sulphur and phosphorus. Nitrogen probably exists in two forms, namely (*i*) as unstable "imino" form that decomposes between 300°C and 900°C, and (*ii*) much more stable "nitride" form. Sulphur is present chiefly in three forms, namely (*i*) pyrites, (*ii*) gypsum, and (*iii*) organic compounds. From lignite to anthracite, there is progressive elimination of water, hydrogen and oxygen and a corresponding enrichment of carbon while nitrogen remains more or less constant. The hydrogen, oxygen and nitrogen constitute the 'volatile matter' in coal. The carbon is mainly present as fixed carbon, though it may also be present in the volatile matter in the form of some compound. Earthy matter is left behind after burning of coal as 'ash', and this ash as well as the moisture, sulphur and phosphorus are regarded as impurities in coal.

The American Society of Testing Materials (A.S.T.M.) in 1937, has prescribed that coal containing 69-86% fixed carbon and 14 to (+) 31% volatile matter (both on 'dry and ash-free', i.e., 'unit coal' basis) should be considered as bituminous coal as distinct from the high-fixed-carbon low-volatile-matter anthracite or the low-fixed-carbon high-volatile-matter sub-bituminous coal and lignite. As per the classification of the Bureau of Indian Standards (B.I.S.), the Indian bituminous coal may contain up to 50% ash (beyond this, the material is regarded as carbonaceous shale), while the volatile matter may exceed 35% (on unit coal basis) and the moisture may exceed 2% (on air-dried basis).

So far as the ultimate composition of hydrogen and oxygen are concerned, their percentages in the bituminous coal are small compared to that of fixed carbon or ash, being generally of the order of 5-6% and 3-12% respectively.

2. Thermal Value

Coal is valued as a source of heat, and the thermal or calorific value is a measure of the usefulness of coal. Heat is produced due to burning of both the fixed carbon and the combustible constituents (e.g., hydrogen) in the volatile matter. The heat value is expressed in either

British Thermal Unit (B.T.U. or B.Th.U) or kilocalories (1 kcal. = 3.9683 BTU or 1 BTU = 0.252 kcal).

The 'gross calorific value' is the total amount of heat obtainable by the combustion of a given coal. Kilocalorie denotes the number of kilograms of water which may be heated through 1°C, in the neighbourhood of 15°C, by the complete combustion of 1 kg of coal. B.T.U. denotes the number of pounds of water which may be heated through 1°F, in the neighbourhood of 60°F, by the complete combustion of 1 lb. of coal. In either of these cases, the conditions are: (*i*) coal is dried at 105°C until its weight becomes constant, (*ii*) whole of heat is transferred without loss to the water, and (*iii*) the products leave the system at the atmospheric temperature and pressure. Gross calorific value is determined in the laboratory with the help of a 'Bomb Calorimeter'. But when the oxygen content in the coal is low, the following empirical formula based on its chemical composition gives an approximate idea of the gross calorific value:

$$Q = [8080C + 34460 (H-1/8\ O) + 2250S]/100$$

Where C, H, O and S are the percentages of carbon, hydrogen, oxygen and sulphur respectively in the dry coal.

In contrast to gross calorific value, the net calorific value does not take into account the heat liberated by condensation of the steam produced on combustion and the subsequent cooling of this condensed steam to water down to atmospheric temperature (15°C or 60°F).

However, there is a peculiar point connected with the utilization of heat developed in burning coal. Sometimes, the calorific value may not be the true indicator of the ability of a mass of coal to generate heat, and it has been observed that coal having lower calorific value may evaporate water better than when it is having higher calorific value. This happens because, sometimes, the volatile matter comes off so rapidly that much of it including the smoke is incompletely burnt. So, in practice, what actually counts, is the 'useful heat value or UHV'. The average UHV of different industrial grades of coal in India generally varies from 3800 to 6640 kcals / kg. However, as per the specifications stipulated by the Government of India in June, 1993, the minimum heat value of the lowest grade of noncoking coal is 1300 k cals / kg.

3. Fuel Ratio

Burning of coal depends on both fixed carbon and volatile matter. While fixed carbon is the steady lasting source of heat, volatile matter causes ready ignition and burns in the form of a gas giving a long smoky flame. In some uses of coal, the relative contribution of both these sources of heat assumes importance. This is determined by the 'fuel ratio', which is the ratio of the fixed carbon content to the volatile matter content. Fuel ratio increases with the rank of coal from lignite to anthracite.

4. Reducing Power

The fixed carbon in coal is a solid reductant. Though hydrogen gas in coal is also a reducing agent, its low content in coal makes its contribution in this regard very insignificant compared to that of the fixed carbon.

The mechanism of reduction by the fixed carbon has been a subject of much controversy in the past. It is now believed that the carbon first becomes incandescent at high temperature. When this incandescent carbon comes in contact with oxygen of air, it forms carbon monoxide

(CO) gas. This CO gas is unstable and has a strong affinity for oxygen. It readily combines with oxygen of the metallic oxide substances and reduces the latter. Also, being a gaseous substance, CO is able to react effectively with solid oxides.

5. Coking Property

Some types of coal, when burnt in absence of air, become plastic at 340-500°C, and then on further heating give rise to a hard, spongy, swollen residue called coke. This characteristic is called coking or 'caking' property, and the coal which yields coke is called coking coal (rarely also 'caking coal'). This property is exclusive to coal, but all types of coal do not possess this property. When noncoking coal is heated similarly, it yields a powdery residue called 'char'. The ability to attain a plastic state and the period during which it remains plastic, are of great significance in the coke formation process. The fusible constituents in coal become plastic and the nonfusible inert ones are dispersed in it, adding to the coherence of the resultant coke. However, it is not clearly understood why some types coal are coking and others are not. It has not been possible to establish a clear correlation between the coking property and any specific physico-chemical or petrographic parameter. The coking characteristic has to be ascertained by trial burning.

Since, during the process of coking, coal is burnt in absence of air, the carbon does not oxidize while the volatile matter escapes. As a result the coke becomes enriched in carbon at the cost of the volatile matter. This increases both the thermal value and the reducing power of the coal. However, besides the chemical properties, the physical structure is also important. The increased strength of coke enables it to withstand the conditions of stress and strain prevailing within a furnace; and its porous structure facilitates the air to permeate and react with the carbon, thus generating heat and CO efficiently and quickly. Coke can also withstand handling and long-distance transportation without generating dust.

The coke is produced in ovens called 'coke ovens'. The more efficient retort type of ovens has replaced the less efficient 'beehives'. In the beehives, some air used to be admitted and as a result, the quality of coke used to be inferior; moreover, complete recovery of the by-products from the volatile matter was not possible. In the retort type of ovens, complete absence of air is ensured with improvement of the quality of coke and efficiency of byproduct recovery.

In India, the best coking coal (prime coking coal) occurs only in Jharia coalfield (in seams numbered X and above), whereas medium or semicoking coal occurs in Raniganj and a few other coalfields. Coking coal containing high sulphur is also known in the states of Assam, Arunachal Pradesh, Meghalaya and Nagaland.

Several laboratory methods have been developed to measure the coking propensities. But they are empirical requiring standard conditions. The most accepted ones amongst them are:

(*i*) Caking index determination.

(*ii*) Low temperature Gray-King assay at 600°C (GKLT).

(*iii*) Swelling index determination.

(*iv*) Gieseler plastometric test.

(*v*) Dilatometric test.

(*vi*) Sapozhnikov test.

Out of these, the first three tests are usually conducted in various countries including

India. None of these tests, however, singly or in combination, yields values that can be correlated with the physical characteristics of the actual coke. At best, these tests may narrow down the selection of samples for trial in pilot plant, which in any case will have to be carried out before deciding the suitability of the coke to any specific use. These tests are described as follows:

(*i*) *Caking index*: This depicts the ability of a coking coal to bind with an inert substance. The test (BIS) consists in carbonizing under specific conditions 25 gms of a mixture of powdered coal and graded sand. The ratio of sand to coal is gradually increased till the cold carbonized button of the mixture (obtained after the test) just sustains a 500 gm weight without crushing. The number of parts of sand per part of coal in the mixture of the limiting strength is called the 'caking index'. Though there is no definite correlation between caking index and physical properties of coke, broadly it has been observed that the caking index of prime coking coal is 20-27, that of medium coking coal around 15-20, and that of semi to weakly coking coal in the range of 10-18.

(*ii*) *G.K.L.T. test*: In this test, 20 gms of powdered coal are heated to 600°C in a silica retort under controlled conditions. The cokes obtained from this test are classified from 'A' to 'G9'; 'A' indicating a incoherent powder at one end, and 'G' strongly swelling at the other. 'G3' and higher types indicate strongly swelling coke. The prime coking coals of India broadly fall in G–G3 class, the medium coking coal in E–G class, and the semi or weakly coking coals in E–F class.

(*iii*) *Swelling index*: This is also denoted as 'crucible swelling number'. For its testing one gram of powdered coal contained in a crucible of standard shape is heated in a gas flame under standard conditions of heating. The shape of coke button obtained is compared with standard profile outlines numbered from 1-9. Range of $4\frac{1}{2} - 9$ is considered to be indicative of good coking coal. In India, the crucible swelling number of prime coking coals is around 3 and that of medium coking coals is around 2.

(*iv*) *Gieseler plastometric test*: In this test, the plasticity of coal is measured. This test is based on the principle that as plasticity of the coal mass increases with rise in temperature, the RPM of a rotating rod with rabble arms dipped in the mass, will also increase.

(*v*) *Dilatometric test*: Dilatometers measure the variation in the length of a column of coal (shaped like a pencil) during heating. During heating, the column of a good coking coal will first contract due to softening and settling down as the pores close in; then on further heating, the coal will start decomposing, and the bubbling gases that cannot escape, tend to raise the top level of the column. This test has been conducted extensively on the coals from Sheffield, UK.

(*vi*) *Sapozhnikov plastometric test*: This test is based on the principle that plastic mass, during carbonization of small laboratory samples which have been subjected to unidirectional and regulated heating from the bottom, registers shrinkage and expansion, and the variation in thickness is measured. The experiment is stopped when the temperature reaches 730°C. This test is widely used in Russia and some East-European countries. Though its applicability to high ash Indian coals (where the plastic layer is too thin) is doubtful, it is believed that a good Indian coking coal should have values of the maximum thickness of plastic layer (MTPL) ranging from 17 to 30 mm.

6. Weatherability

If coal is stacked for a long time in open air, it becomes heated, and if kept stacked for a still longer time, it may catch fire spontaneously. This phenomenon is due to the fact that all coals—some more than others—absorb and slowly combine with oxygen on exposure to air even at ordinary temperature. This atmospheric oxidation or weathering of coals is an exothermic reaction which goes on slowly and continuously whenever they are stored or handled with free access of air. It continues with increasing rapidity as the temperature rises, and if the generated heat is allowed to accumulate, it may ultimately give rise to spontaneous ignition. Even if that stage of spontaneous ignition is prevented, the weathering itself causes loss of heating value of the coal. It is believed that presence of pyrites within coal may also have some role in weatherability of the coal.

This property of weatherability is not directly responsible for any use of coal. But it affects negatively the economics of utilization of coal inasmuch as it requires extra care in handling, transportation and storage of coal, and also it causes some loss of the thermal value of coal.

7. Specific Gravity

The specific gravity of Indian coals varies from 1.3 to 1.7 depending on the contents of various constituents like carbon, volatile matter, ash, sulphur, etc. Use of coal does not directly depend on its specific gravity. However, sometimes the coal needs to be beneficiated or 'washed' by gravity separation method, and then this property becomes relevant.

8. Abrasive Power

Hardness, i.e., resistance to abrasion or scratch, is not of much economic significance in case of coal. But the abrasive effect of coal on other substances is of importance when the coal is pulverized, say, in a ball mill. This abrasive effect depends on grains of pyrite, sand, etc., which may be present within the coal. High content of such abrasive grains may damage the balls.

9. Grindability

This is the combined manifestation of a number of physico-mechanical properties like toughness, hardness, strength, etc. This is of economic significance when the coal is required to be pulverized. Coal of poor grindability will require more power for pulverization. There are various tests to measure the degree of grindability. But the basic principle is to relate the power consumption to the increase in surface area of the coal (because more the coal is ground, more is the number of grains generated, and so more is the total surface area).

10. Ash Fusion Temperature

In some of the smelting technologies, non-coking coal containing relatively high ash is charged into the furnace. In such cases the fusion temperature of the ash affects the temperature at which the coal starts softening, and the efficiency and economics of the operation. This temperature depends on the composition of the ash (relative percentages of alumina, silica, etc). In Indian non-coking coal, this temperature may be of the order of 1200°C or more.

11. Gamma Ray Absorption

Coal absorbs less gamma ray than the mineral matter residing within it. Consequently, the gamma rays transmitted through coal are attenuated in a lesser degree compared to those transmitted through non-coal matter like shale, stone, etc.

USES

Coal has entered our life so much that it is difficult to make an exhaustive list of the actual uses to which it is put both directly and indirectly. However, following is a list of the important and more or less direct uses of coal.

1. Beneficiation.
2. Smelting of iron ore and other oxide ores for manufacturing pig iron, sponge iron and other metals.
3. Extraction of chemical products:
   - (*a*) Ammonia
   - (*b*) Benzole
   - (*c*) Benzene
   - (*d*) Aniline
   - (*e*) Toluene
   - (*f*) Xylene
   - (*g*) Naptha
   - (*h*) Pyridine
   - (*i*) Creosote oil
   - (*j*) Naphthalene oil and naphthalene
   - (*k*) Carbolic acid
   - (*l*) Cresols
   - (*m*) Xylenols
   - (*n*) Phenol
   - (*o*) Anthracene oil
   - (*p*) Pitch
4. Generation of gas:
   - (*a*) Coal gas
   - (*b*) Producer gas
   - (*c*) Water gas
   - (*d*) Carburetted water gas
5. Fertilizers
6. Domestic heating
7. Cement manufacturing
8. Brick burning

9. Power generation
10. Locomotives
11. Synthetic petroleum
12. Calcium carbide
13. Activated carbon
14. Chloro-fluoro carbon (CFC)
15. Foundry
16. Sialon ceramics

SPECIFICATIONS OF USE

Nature has not created coal everywhere according to the specifications demanded by man and industry. And yet, in India, it is occurring abundantly. So, we find that in many uses either compromises are made or some degree of preparation or processing of the raw coal is practised. The various desirable specifications, compromises and innovations of different uses are discussed as follows.

1. Coal Beneficiation

In case of coal, beneficiation is traditionally referred to as 'washing', because conventional processes are all 'wet'. Beneficiation is not an end-use of coal. Coal concentrate is an intermediate product. For many end-uses of both coking and noncoking coal, higher fixed carbon and lower ash content than what nature has endowed, is desirable from both economic and environmental standpoints. Economic considerations include lower transportation cost per unit heat value, better handling and operational efficiency in the consuming plant, lower maintenance cost and longer life of the plant, etc. Environmental considerations mainly result from accumulation of waste ash at plant sites and other unwanted locations. The processes of coal beneficiation can be classified as follows:

A. Wet process
- — Dense medium
- — Natural medium
  - (*i*) Barrel technology
  - (*ii*) Cyclone technology
    - (*a*) Water-only cyclone
    - (*b*) Hydrolyzer
    - (*c*) Gravity spiral
    - (*d*) Froth flotation
  - (*iii*) Jigging

B. Dry process
- — Mechanized sorting
- — Photometry sorting
- — Conductivity sorting

— Microwave processing

— Gamma radiolytic process

— Pneumatic tabling

These methods require grinding of coal for liberating the mineral particles residing within it, grindability and abrasive power invariably become very important criteria in use of coal for beneficiation. The dense medium technique is based on differences in the specific gravities of the coal particles (lighter) and the mineral matter particles (heavier). So specific gravity of coal is also relevant.

Particles of coal having sizes above 10 mm are subjected to dense media separation. The specific gravity of the medium is suitably increased by adding magnetite grains, and it is so adjusted as to enable the coal particles to float and the mineral particles to sink. Separation takes place in a slowly revolving (1-3 rpm) horizontal drum; and it is achieved entirely by the buoyancy of the medium, the dynamic effect playing no part.

In the natural medium processes, only water is used as the medium. The water combines with the fine coal and shale particles to form a viscous natural medium. In the barrel technology, an internally scrolled rotating (5-20 rpm) downward tilted (8° angle) barrel is deployed. The combination of viscosity of the medium and the dynamic effect of the barrel (produced by its rotation and the movement of the feed material due to gravity because of the tilt of the barrel) causes the coal to float. In the cyclone, the small particles are separated by centrifugal and vortex action. In the jigging technique, vertical up and down or pulsating movement of the water is created by air blown into the compartments of the jig shell. The pulse results in a high degree of bed fluidization that allows free movement of particles in the bed, and the particles settle in stratified manner with heavier fractions settling towards the bottom layers.

In sorting, the size of the coal should be fairly coarse—usually greater than 10 mm—so as to facilitate distinguishing coal from non-coal matter on the basis of visual assessment of specific physical properties like colour. In mechanized sorting, unlike in pure hand sorting, the sorters send signals to some mechanical or electronic device which separate the particles of different matter. In photometry sorting, the sorting is done, instead of by hand, by a photometry sorter which works on the principle of scanning laser light reflected differentially from coal and non-coal matter and computerized control system for releasing air blasts to throw out selected particles. Generally the device is effective for particles ranging in size from 10-150 mm. In conductivity sorting, the differential electrical conductivity of the coal and non-coal particles is the basis for sorting. The preferred size range is 50-150 mm.

Since the early 21st century, a new dry beneficiation technology has been undergoing trials. The technology makes use of grindability of coal containing mineral matter, differences in specific gravity and gamma ray absorption between the coal and mineral matter particles. Both specific gravity and gamma ray absorption are lower in coal than in mineral matter. Ground particles on a conveyor belt pass between an emitter system emitting gamma rays and laser beams and a detection system to detect the intensities of the signals transmitted by the coal and the mineral particles. On the basis of the differences in the intensities of the transmitted signals, a computer calculates the specific gravities of the particles and signals a pneumatic system to blow compressed air pointedly and selectively directed to the heavier mineral matter particles, which are thrown further away than the coal particles which are left out by the compressed air beams. So, in this technology, besides grindability the other properties that become important are abrasive power, specific gravity and the property of coal relating to gamma ray absorption.

The grindability and abrasive power may not, however, be of any relevance if the coal could be cleaned in lump form. In the mid-1980's, in U.S.A., a novel process using microwave was developed. This microwave coal clean-up process was based on the ability of microwave beams (tuned to the proper frequency) to pass through the coal and strike the mineral particles within it, which absorbed and converted the radiation into heat. As a result, a high temperature developed deep within the coal mass, causing decomposition of the mineral particles into water-soluble substances that could be easily washed out. However, this technology has not found commercial application.

In the gamma radiolytic process, being experimented in the Central Fuel Research Institute, India, during early 21st century, aqueous/acidic coal slurries are irradiated by high energy and ionizing gamma rays from cobalt-60 for removal of organo-sulphur by oxidation.

The Government of India, in June 1993, has stipulated that Indian coking coal containing (+)18-35% ash should be regarded as 'washery grade'. In actual practice, the coking coal washeries in India wash raw coal containing 24-33% ash producing concentrates containing 19-22% ash, coal containing over 50% ash going as tailings.

However, the technology and necessities have undergone considerable changes during the subsequent decade. Now non-coking coal is also beneficiated and the ash content in non-coking coals of India can go much higher-up to 50 per cent.

2. Pig Iron Manufacturing

By far the most important use of coal is in smelting of iron ore to make pig iron in conventional blast furnaces. In large blast furnaces, the charged material is required to withstand the very high stress exerted by the ascending hot furnace gases, and also to withstand the stress of very rough handling and bulk charging into the blast furnaces. If under these stresses the material crumbles and pulverizes, then it will tend to block the passage of the gases. So, the coal that is charged into a blast furnace, must have the requisite high physico-mechanical strength, and yet be porous enough to allow air and gases to pass through. For rendering this combination of strength and porosity, coal has to be converted to coke through high temperature carbonization for use in iron smelting in blast furnace.

The function of coal in pig iron manufacturing is three-fold : (*i*) to provide heat, (*ii*) to provide reducing gas and (*iii*) to act as a solid reductant. The heating value and the reducing capability are provided by the carbon. So the carbon content must be very high. For enhancing the carbon content in the raw coal, the latter should be converted into coke for use in blast furnace. The reducing gas is generated due to reaction of the carbon with the oxygen of air, and for efficiency of this reaction, a large surface area of the carbon should be exposed to the reaction. Here again the porosity and high carbon content of the coke provide the answer. To sum up, the coal must be used in the form of coke, and so the most important specification of coal for use in pig iron manufacturing is that it must possess good coking property.

An important criterion in pig iron production is the output per unit volume of the blast furnace. In the early 1940s, average productivity of Indian blast furnaces was 0.97 tonne per cubic meter a day, compared to 0.84 in the erstwhile USSR and 0.76 in Japan. In the 1970s, the productivity in U.S.S.R. and Japan reached 2.0-2.5 tonnes per cubic meter a day, in India it has hardly improved. The differences in technology and alumina content of the iron ore apart, that in the ash of the coal has also contributed to this situation. The Indian coal used in iron smelting contains as high as 23% ash compared to 10-13% or even less in the Western countries. Due to these factors, the consumption of coke per tonne of pig iron, i.e., the coke rate in blast furnaces is high—of the order of 750-800 Kgs compared to about 375 Kgs in Japan. High ash content reduces the availability of carbon in the coke, thus requiring use of increased quantity of coke. This in its turn increases the already large quantity of ash generated within the blast furnace that adds to the volume of slag formed and eventually decreases its effective space. So the second most important specification of coal for use in pig iron smelting is that its ash content should be low.

The low ash content in the coking coal of the Western countries can yield coke containing less than 10% ash and as high as 90% fixed carbon, and this largely contributes to the low coke rate and high productivity of the blast furnaces. In India, on the other hand, the high ash content in the natural coking coal necessitates a complicated balancing between the cost of upgrading the coal and the efficiency of the blast furnace. Nevertheless, some degree of upgradation (or washing) of coal is practised. As per the stipulation of the Government of India in June 1993, coking coal containing up to 18% ash should be regarded as 'steel grade'. However, in practice these specifications are relaxed due to the exigencies of availability, and through washing, the ash content of the coking coals is reduced from over 24% to about 18-22% (in Western countries also, though the ash content is lower than in India, washing is practised to the extent that cost of washing remains less than the cost saved through increased blast furnace efficiency). This washed coal yields coke containing 24-26% ash and 70-73% fixed carbon. Some imported low ash coking coal may be blended with the washed coal to bring down the ash content in the coke feed further to 17% or so. Thus it can be seen that the ash content specified by the Indian iron industry depends on a number of interrelated factors like cost of washing, cost of imported coal, cost of transportation, cost of smelting iron, the Al_2O_3/SiO_2 ratio in the iron ore, that in the coke ash and that in the slag, etc. The net outcome should be a positive return on investment. Added to this is the need for extending the life of the coking coal reserve.

During the late 20th century, a new technology involving 'coal dust injection or CDI' has been developed. In this, non-coking coal is accepted by the blast furnace. But here again ash content of the coal should be low, and in addition, the ash fusion temperature should also be low. The non-coking coal generally preferred by the Indian industry contains 16% maximum ash, 4% maximum moisture (at 40°C and 60% RH), and 25-30% volatile matter (on dry basis). The point of first softening should be at around 1200-1250°C.

3. Sponge Iron Manufacturing

Sponge iron (broadly it includes hot briquetted iron and hot metal also) is an alternative to pig iron as a raw material for steel. It is a porous lumpy mass of almost metallic iron, that is obtained by direct reduction of iron ore in the solid state, i.e., without the necessity to melt the ore. This technology of making sponge iron is therefore also known generally as 'direct reduction technology'. It can also be made by various other processes like Corex, Dios, Romelt, etc. that fall under a sophisticated version of this technology (smelting reduction technology). The sponge iron mixed with iron scrap is then charged into an electric arc furnace (EAF) and melted for manufacturing steel. The direct reduction of iron ore to sponge iron can be effected with the help of either natural gas or coal. While the gas-based technology was first developed in U.S.A., the Lurgi of Germany was the pioneer in developing the coal-based technology. This technology is suited to mini-steel plants.

In this technology, for reduction of iron ore, non-coking coal can be used, because:

(*i*) the reduction takes place in comparatively low temperature and so a high content of carbon like that in coke is not essential;

(*ii*) sponge iron is manufactured in horizontal rotary kilns where the coal is not subjected to a high degree of stress as in large-sized blast furnaces.

Further, since the reduction takes place in solid state, the ash of the coal does not enter into the sponge iron. The purity of the sponge iron (i.e., the degree of metallization of the iron ore) is therefore not affected by the ash content in the coal. Thus a low ash coal is also not essential. The only consideration is that the fused ash is deposited on the inner wall of the kiln, and so more ash means quicker deposition and more frequent cleaning and shut down. In India, ash content up to 24% is preferred by the industry (except in Corex process in which the preferred ash content is 5-12%); but in reality, the industries accept beneficiated coal containing up to 29% ash or even raw coal containing up to 45% ash. The other specifications as preferred by the industry, are as follows:

(*i*) *Direct Reduction Technology*: Volatile matter 23-34% (on dry basis); Sulphur 1.0 % maximum; Inherent moisture 8.0% (preferable), 11% (maximum); Caking Index < 5 (non-coking coal); Size (–) 20 mm; Initial softening 1150-1200°C minimum, but 1250°C desirable.

(*ii*) *Corex*: Non-coking coal containing 60-75% fixed carbon, 20-35% volatile matter, moisture 3-6%, sulphur 0.4-0.6%, and phosphorus 0.2% maximum. Size range 5-40 mm. (preferred) and 0-50 mm (acceptable).

(*iii*) *Dios*: High volatile matter up to 40%; size 1 mm maximum.

(*iv*) *Romelt*: Moisture 10% maximum; volatile matter up to 20% desirable, but up to 40% tolerable.

However, in large size plants (but nevertheless much smaller than conventional blast furnace) the strength, heat value and reducing power of the coal is enhanced to some extent. For this purpose, the non-coking coal may be converted to 'char' (or soft coke) by subjecting it to low temperature carbonization at 450-700°C temperature. Char contains about 9-20% volatile matter, the hydrogen of which takes part in the reduction process. The increased carbon content in the char increases the thermal value and the reducing power.

4. Smelting of Other Oxides

For producing elemental phosphorus from its oxide, metallic copper from its oxide ore, ferromanganese from manganese ore, etc., hard coke is used as in the case of pig iron manufacturing in blast furnace. So, the coal should possess good coking property and low ash.

5. Extraction of Chemical Products

The volatile matter in coal is composed of hydrogen, oxygen, nitrogen and carbon (this carbon is in addition to the 'fixed carbon' of coal). This volatile matter is the source of various organic chemical substances. For extracting them, the coal is first carbonized, i.e., heated in absence of air. This makes the volatile matter come out in the form of smoke without burning. When this smoke is cooled, one part of it escapes as gas while the other part condenses and settles down in two layers of liquid substances—the lower coal tar and the upper aqueous liqueur. It is these liquid substances that contain the various chemicals. The aqueous liqueur contains ammonia and ammonium salts—formed by the nitrogen and hydrogen of the volatile matter. The tar contains a host of complex hydrocarbons, and on fractional distillation it yields different derivatives.

The chemical derivatives are used for manufacturing many products. Ammonia can be used to manufacture soaps, fertilizers, detergents and ammonia liqueur. Benzole can be used as motor benzole, while benzene is a raw material for making DDT, nylon, aniline dyes and scent; toluene for TNT and saccharine; xylene for paints, varnishes and printing ink. In the light-oils group, naptha is used in brake linings and lino; pyridine in photochemicals; and creosote oil in aviation fuel, fuel oil and timber preservative. Naphthalene oil is used in plastic, fire fighting chemicals and moth balls. Carbolic acid yields cresol and xylenol—the former finding use in tanning and weed-killing substances, and the latter in antiseptics and disinfectants. Phenol is a raw material for adhesives and aspirin. In the higher distillate group, the two most important fractions are pitch and anthracene oil. Pitch is used in roof coating and rust preventive substances, while anthracene oil finds use in road tar and fruit tree sprays. But these derivatives are not to end in themselves, as technology is advancing, it may be possible to further differentiate these fractions to yield more and more compounds. For example, in the pitch fraction itself, as many as 5000 compounds have been estimated to be present, out of which hardly 75 have actually been separated. In the 1990's, the National Physical Laboratory (NPL), India developed processes for two types of coal tar pitches namely, performing pitch and impregnating pitch for use in carbon-carbon composites for the defence applications, and it was claimed to have potential applications in the manufacture of graphite electrodes, needle coke, carbon fibers and high density isotropic graphite.

It is obvious that the chemical derivatives of coal depend on its volatile matter content. It is also known that the relative quantities of the different derivatives are determined by the degree and technology of carbonization. However, it has not been clear whether coking property of the coal bears any direct relationship with the types and quantities of the products. Indirectly, however, it does have some influence as follows:

- Firstly, these coal tar derivatives are recovered as byproducts during manufacturing of hard coke through high temperature carbonization of coking coal or during manufacturing of char or soft coke through low temperature carbonization of non-coking coal. Though, coal was used in Germany exclusively for obtaining tar by low temperature carbonization during World War-I and World War-II, those tars were utilized solely as petroleum substitute, and not appreciably for extraction of chemicals. Since out of the two principal products namely hard coke and char, the former fetches much higher price in the market, it is economically more advantageous to extract the chemicals from coking coal than non-coking coal.
- Secondly, during high temperature carbonization of coking coal, practically the whole of the volatile matter content is extracted, while in case of low temperature carbonization of non-coking coal, the extraction is partial. So, from the sole point of view of recovery of chemicals also, the coking coals are more productive than the non-coking coals.
- Thirdly, the range of variation in the chemical composition of non-coking coal is wider than in that of coking coal. As a result, the nature of the chemical products

derived from coking coal is more uniform and predictable. This factor also justifies the preference shown by the industry in favour of coking coal.

However, as such, there is no industrial specification for this end-use of coal.

6. Generation of Coal Gas

As has been mentioned, during carbonization of coal, while one part of the distillate condenses on cooling, the other part escapes as gas. This gas is known as coal gas. On an average, this gas has a heating value of 144.4 kcals. This gas is valued for both illuminating and heating power. In the earlier times, coal gas was extensively used for street lighting purpose. Even now, it is used in Davy's Lamp for lighting purpose inside coal mines. It is also used in Bunsen burner and other similar burners for giving smokeless flame. The luminosity of coal gas is believed to be due to the decomposition of part of the gas and deposition of solid carbon towards the interior of the flame, the incandescence of those carbon particles contributing to emission of light. This explains why gases rich in heavy hydrocarbons (ethane, ethylene, propylene, benzene, etc.) produce more luminosity. This is so because these readily decompose under the influence of heat, and produce carbon. On the other hand, constituents like methane and hydrogen yield negligible or nil carbon and so do not produce a very luminous flame.

Researchers are now believing that through gasification of coal, hydrogen can be produced as a by-product and that it will in future emerge as the next primary fuel source. They are also believing that these hydrogen-rich gases could be used for power generation, in fuel cells (for details, see the chapter on Natural Gas), as liquid fuels or for chemicals production. According to International Energy Agency (IEA), in 2003, some 1800 MW was being generated in plants based on integrated combined cycle (IGCC) systems using gasified coal and another 3150 MW was being planned.

As in the case of chemical products, the quality of coal gas also depends on the chemical composition of the volatile matter in coal.

7. Generation of Water Gas

Water gas (also called 'blue water gas') is produced by passing water or steam over coal or coke surface heated to over 1000°C. The hot incandescent carbon and the steam interact with each other and yield a mixture of hydrogen and carbon monoxide as represented by the equation:

$$C + H_2O = CO + H_2$$

This reaction is endothermic and the absorbed heat is stored in the water gas, the thermal value of which is on an average 75.6 kcals. This gas is used in steel welding operations.

In this case the carbon in coal is gasified completely, and the quantity of gas generated will depend on the carbon content of the coal. If coke is used, more carbon will be available and more gas will be generated.

8. Generation of Producer Gas

This gas is produced by passing air over red hot incandescent coal bed. This is a low grade fuel possessing on an average 37.8 kcals heat value. This can be produced very cheaply and its chief use is in firing of industrial furnaces (e.g., glass-making furnace, steel-making furnace) and also for starting large diesel generators used in factories. Any coal including low-grade high-ash ones, can be used for generation of producer gas.

Sometimes, instead of air, a mixture of air and steam is used. It enables up to 75% of the nitrogen of the coal to be converted into ammonia (NH_3) which can be recovered as ammonium salts.

9. Carbureted Water Gas

A mixture of 80% coal gas and 20% water gas is sometimes used. When this is further enriched with gaseous hydrocarbons obtained through cracking of petroleum, it is known as "carbureted water gas" (heat value 138.6 kcals).

10. Fertilizers

The nitrogen content of the volatile matter in coal can be utilized for making ammonium sulphate [$(NH_4)_2SO_4$], which is a nitrogenous fertilizer. For manufacturing this, ammonia (NH_3) is first recovered from coal and then reacted with sulphuric acid (H_2SO_4) according to the equation:

$$2NH_3 + H_2SO_4 = (NH_4)_2SO_4$$

There are two ways of recovering ammonia from coal.

In the first method, it is extracted as one of the byproducts, during carbonization of coal. As has been mentioned earlier, the process of carbonization yields a residue of coke or char, while part of the volatile matter coming out, when condensed, settles down in two layers—the lower tar and the upper ammonium liqueur. But still, some ammonia remains in the tar and can be recovered by fractional distillation. In this process both nitrogen and hydrogen are contributed by the volatile matter, and the amount of ammonia recovered is limited by the hydrogen content of the coal, which is much less compared to the nitrogen content. This method of recovering ammonia suffers from the disadvantage that considerable portion of the nitrogen of the coal remains unutilized, and also that it depends on the production of the principal product—namely coke or char, in order to be cost-effective.

The second method consists in complete gasification of coal dust. As has been mentioned earlier, a mixture of air and steam blown over hot incandescent coal bed yields producer gas and also ammonia. The latter can then be recovered as ammonium sulphate. In this method, additional hydrogen is provided through the steam which reacts with the nitrogen of the coal. Thus 65-75% of the nitrogen available in the coal can be converted into ammonia. This method is cheap and any type of coal including low grade fines of non-coking coal can be used.

11. Domestic Heating

Coal is used in a large number of houses for cooking, heating, etc. The most important specifications are:

(*i*) Coal should give out as little smoke as possible.

(*ii*) It should catch fire quickly.

(*iii*) It should produce steady and lasting heat.

Smoke is a nuisance in any house and is injurious to health. The smoke that is generated due to burning and oxidation of carbon, is unavoidable (otherwise heat will not be generated). But attempts are made to minimize the smoke produced due to burning of the volatile matter. At the same time, since it is the volatile matter that ignites first and produces the necessary fire to initiate the process of burning of the carbon in coal, too low a volatile matter content is not desirable. Due to this reason, hard coke or anthracite prevents the coal from catching fire quickly. In domestic heating, this problem assumes added importance, because small quantities of coal are required to be fired repeatedly at short intervals. One way is to use soft coke in which the volatile matter is partially driven out through low temperature carbonization. But in this case, though some chemicals can be recovered, still the soft coke becomes too costly to be afforded by the common mass—particularly in a developing country like India. A cheaper process more commonly followed in India is to heap up the run-of-mine coal in the form of beehives and to set fire to them; when most of the smoke goes out, the fire is extinguished by sprinkling water.

Besides an optimum amount of volatile matter, carbon content is also important. It is the carbon that produces the effective and lasting heat. However, cooking etc. does not take very long time and hence, a very long-lasting heat (and hence a very high carbon content) is not required for most of the domestic purposes. To sum up, the fuel ratio of coal should neither be too high nor too low.

In ordinary houses, coal is used in small quantities for relatively short durations of time, and hence for each operation, the quantity of ash produced is limited so as not to pose any disposal problem. So, the ash-content in coal is not of much consequence as such, except that too high ash content will proportionately decrease the carbon content (and hence the heat value) of the coal.

12. Cement Manufacturing

In cement industry, coal is used to provide the necessary heat for the reactions to take place within the furnace. Coal is charged into the furnace along with other raw materials, so that the entire charge of the raw materials is evenly subjected to the heat generated. Therefore, coal may be of the non-coking type, but it should have high thermal value.

The ash content, however, should be as low as possible, because after burning out of the coal inside the furnace, the SiO_2 of the residual ash gets into the reactions and ultimately, into the cement product. If the ash content of the coal is high, then low-silica limestone has to be used. So, in effect, the ash content is specified according to the cost-benefit ratio of using low-ash coal or low-silica limestone.

The ash content is also specified according to the type of kiln used for manufacturing cement. In case of conventional kilns, coal and other raw materials are charged in a loose mixture. About 75% of the ash is blown off, and only the remaining 25% enter into the clinker. So, this technology can accommodate coal with a relatively higher ash content. Such industries in India now-a-days are reported to accept beneficiated coal containing up to 32% ash. On the other hand, mini-cement plants employ vertical shaft kilns. In this process, coal is mixed with other raw materials, pelletized, and then the pellets are charged into the kiln. So, there is no chance of any ash to be blown off, and the entire ash goes into the clinker. This technology, therefore, requires very low content of ash in coal, unless low-silica limestone is used.

13. Brick Burning

In brick burning, a steady lasting heat is required for ensuring slow but uniform heating of the raw bricks. Quick and sudden rise in temperature may cause cracks in the bricks. So, coal with high thermal value but low volatile matter content is preferred. Since coke is too costly compared to the price of the bricks and also initial firing is difficult due to practically total absence of volatile matter, char or soft coke made out of non-coking coal is preferable in brick burning industry. Some quantity (around 10%) of volatile matter facilitates initial firing, and thereafter, a reasonably steady heat can be maintained. Unlike in the case of domestic heating, frequent firing at short intervals is not required and also, the price of bricks can justify the cost of char.

The ash has no chance of getting into the composition of the bricks, and so a high content as such is not objectionable, provided the thermal efficiency of coal is not impaired too much.

14. Power Generation

The principle lies in conversion of the heat energy of coal into mechanical energy. In modern thermal power generation plants, coal is used to heat water in boilers, transform the water into superheated steam and then direct the steam at great force for moving turbines. In the most prevalent practice in India, coal, after primary crushing, is pulverized into micron size in ball mills and then mixed and transported with compressed air to the firing system of the boiler. This technology is called 'pulverized fuel combustion (or PFC)'. The most important features are that the steam has to be generated round the clock and the rate of burning of coal should keep pace with the rate of evaporation of water.

Volatile matter content of the coal is important. Too high a volatile matter content may burn the coal more rapidly than the rate of evaporation of water, and consequently, considerable heat value may be lost because: (*i*) some carbon may remain unburnt, and (*ii*) portion of the heat generated may not be transferred to the water. On the other hand, too low a volatile matter content will slow down the rate of generation of steam. A range of 16-23% volatile matter (on dry and ash free basis) may be regarded as ideal. Further, the non-coking variety of coal is used because coking coal is scarce and costly.

Since the coal is required solely for generating heat (and not for any chemical reaction), a high thermal value is obviously desirable. For this purpose, a high carbon content is required.

In thermal power plants, large volumes of coal are burnt round the clock. So, a high ash content will result, over a period of time, in accumulation of too large a quantity of reject to be easily disposed of. Moreover, if the power plant is situated far away from the source of coal, then transportation of coal containing high ash (and hence correspondingly lower fixed carbon) will mean a high cost of transportation per unit heat value. The ash as such does not directly affect the generation of power but for this problem of disposal and higher effective cost. In India, power grade non-coking coal (or 'steam coal' as it is called) with ash content between 30 and 50% is supplied to the power plants. It all depends on the design of the boilers, the economics and the legislative stipulations. The Government of India has stipulated that in thermal power plants located at distances of 1000 km or more from the source of coal or at otherwise environmentally sensitive areas, the ash content in the coal—beneficiated or raw—must not exceed 34 per cent.

Sulphur should be very low, because, in view of the large volumes of coal to be burnt continuously, even a small percentage present in the coal will result in emission of significant quantities of sulphur into the atmosphere and that may cause air pollution.

15. Locomotives

This is a direct use of coal in transportation and is fast receding into obsolescence. In this case also, coal is used for generating steam in boilers. The main differences between steam generation in a large power plant and in a locomotive engine are:

(*i*) much smaller quantities of coal are burnt at a time in locomotives;

(*ii*) disposal of wastes are spread over much larger geographic areas, in case of locomotives;

(*iii*) in case of a locomotive, it is more important for a charge of fuel to sustain for long duration and not to burn out causing stoppage of the locomotive in the middle of its journey.

These characteristics require that volatile matter of coal neither be too high nor too low (for the same reasons as in the case of power generation). But carbon should be particularly high so that a small quantity of coal may burn long enough.

The ash content does not pose much of a disposal problem (unlike in the case of a power plant), because of small quantities and of distribution of the waste amongst different points along the track. However, a high ash content will unnecessarily increase the quantity of coal to be carried, so a low-ash coal is preferable.

Sulphur is objectionable because of its potentiality as an air-pollutant.

16. Synthetic Petroleum

Germany was the pioneer country to produce economic quantities of oil from coal. In U.S.A., Russia, France, U.K., South Africa, etc., considerable research has been carried out for using coal for the production of synthetic petroleum. At present, in South Africa, commercial plants are in operation. In India also, some experiments have been conducted in this area.

The production of coal-based liquid hydrocarbon essentially consists in hydrogenation of coal or coal-based products like tar and creosote oil. This hydrogenation is more efficient if carried in two stages instead of one. In the first stage, coal or tar or oil is hydrogenated to yield a middle oil, and then the latter is again hydrogenated under high pressure (300-700 atmospheres) with a fixed bed of catalyst. The high pressure automatically helps in obtaining a higher temperature. Various catalysts have been tried. These are: Mo, WS_2, WS_2 on HF-activated fuller's earth, Fe on HF-activated fuller's earth, Ni on silica-alumina, cobalt molybdate on activated alumina, tin oxalate and NH_4Cl mix, red mud and $FeSO_4.7H_2O$ mix, $FeSO_4$-caustic soda-Na_2S mix supported on activated carbon, bog iron ore and sulphur mix, etc. The catalysts are usually not recyclable, and iron catalysts enjoy a cost advantage over other catalysts. The composition of the catalyst is the key to the efficiency of the process, and has been a subject of much research.

The coal considered suitable for manufacturing synthetic petroleum should be non-coking. The coking coal tends to become plastic on heating. But for effective mixing of catalysts and for reactions to take place efficiently, it is necessary that the coal is in the form of grains. One of the reasons for applying high pressure during hydrogenation is to prevent the danger of coking (coke is formed in vacuum, i.e., very low pressure).

High-carbon and low-ash coal is suitable. It is the reaction of carbon and hydrogen to form methane that, to a great extent, determines the yield of liquid hydrocarbons. Some of the bituminous coals successfully tried in Germany contained over 80% fixed carbon (dry and ash-free basis) and 3-7% ash (dry basis).

It is believed that suitability of coal for hydrogenation process improves with increasing hydrogen content. It has been observed that for the same carbon content, coals with higher hydrogen content have yielded less gaseous hydrocarbon and more liquid hydrocarbon than those with lower hydrogen content.

17. Calcium Carbide

A mixture of quick lime (CaO) and coke when melted in an electric furnace, yields CaC_2, i.e., calcium carbide. Instead of coke, char may be used. In this case, coal is used not for its thermal value but for its carbon content. If char is used, then the coal can be of non-coking type.

18. Activated Carbon

Activated granular coal is used as a cation exchanger for hard water treatment. Coal can be activated by treatment with concentrated sulphuric acid followed by washing with dilute NaOH. The interaction between H_2SO_4 and coal results in carboxyl and sulphonyl concentrations associated with ion exchange property. Catalysts like mercurious oxide may help the process.

19. Chloro-fluoro Carbon (CFC)

Of the products produced by halogenation of coal, CFC is the only one to be of commercial interest. This may be in the form of a transparent fusible solid or oil or gas, and can be produced by fluorination of low rank coal. CFC is extensively used as a refrigerating agent.

20. Foundry

In this industry, pure metal is melted and cast into different shapes. Here coal is used only for its thermal value. So, very high carbon content is desirable. It is imperative that at this stage no fresh impurity should get into the molten metal. Hence a very low ash content in the coal is also specified, and not more than 2-3% is desirable. Further, the smoke due to volatile matter creates operational problem. Combination of all these parameters point to a very low-ash coke, which can be obtained only from a low-ash coking coal. In India neither coal nor coke of such specifications is produced, and the industry prefers imported coke.

21. Sialon Ceramics

It is an advance material comprising a mixture of silicon, aluminium, oxygen, and nitrogen (i.e., Si-Al-O-N). Sialon is suitable for applications requiring high mechanical strength at elevated temperatures, high specific strength (for weight saving without sacrificing strength), high hardness and toughness, low coefficient of friction and good thermal shock resistance. Possible uses may include refractory brick or material for resisting molten metal, heat engines welding shrouds, gas turbine engines, metal cutting, etc. Ordinary sialon can be made by reacting a mixture of clay and coal in a nitrogen atmosphere.

UTILIZATION OF WASTES

Various kinds of wastes in coal industry may be encountered. These are discussed as follows:

1. Natural Wastes

These include the coal which is either not minable or not usable or both. Generally these kind of *in situ* wastes are accounted for by very deep-seated coal beds and high-sulphur coal.

(a) *Underground gasification*: Mining of coal in solid form from deep-seated coal beds is not economically viable. But it is possible to recover the energy and the chemical values of coal. Russia is the pioneering country in developing methods of underground gasification of coal through intensive research during 1930s after the end of the World War-II, i.e., after 1944, Belgium, USA, France, Poland and U.K. followed suit. Various methods of *in situ* gasification of coal are known. Out of them, the 'percolation' or 'filtration' method is the one that may be deployed for deep seated deposits which are not amenable to development through shafts and galleries. Essentially, the method involves drilling of two or more boreholes, establishing connecting paths within the coal bed and creating cracks and fissures within the coal deposit to make it permeable to gas. A set of boreholes is used as the inlet system through which air or oxygen is pumped and ignition at the base of the borehole is effected. Another set of boreholes is used as the outlet system through which product gases containing thermal and chemical value, come out. For connecting these two systems of boreholes, two methods can be deployed :

(*i*) electrolinking-electrocarbonization and

(*ii*) hydraulic or pneumatic fracturing.

In India, as at the beginning of 21st century, underground gasification of coal is yet to be practised on a commercial scale.

(b) *Coal bed methane*: Coal bed methane or CBM is formed during coalification, the process that transforms plant material into coal. Organic matter accumulates in swamps as lush vegetation dies and decays. Over time, sediments are deposited on the decayed organic matter. As the thickness of the overlying sediment increases, so does the temperature. This creates physical and chemical changes in the organic matter, resulting in the formation of coal and the production of methane, carbon di-oxide, nitrogen and water. As heat and pressure increase, the carbon content or rank of the coal increases, and generally, as has been seen, with increase in rank and depth of the coal seam, its entrapped methane content becomes higher. Coal beds generally do not release this methane to the atmosphere unless produced by a well, exposed by erosion or disturbed by mining. Now-a-days, only the methane extractable by drilling wells is referred to as CBM.

CBM extraction from deep-seated coal beds, otherwise not economically minable, is a recent technology in India being put into practice at beginning of the 21st century, when exploratory investigations have started. In U.S.A., it was extracted commercially first in the 1980's, followed by Australia and China. In U.S.A., as in 2003, CBM production was at the level of 43 billion cubic meters accounting for about 8% of the country's annual gas production. Unlike natural gas, CBM is adsorbed in highly permeable coal beds because of large internal surface area available in the pores. The adsorption is more if the overburden is very thick and the coal bed is under very high pressure (which is the case with deep-seated beds). It is clean and environment-friendly energy mineral like natural gas, with potentiality for use in power generation, fertilizer manufacturing, cooking gas, etc. In India, the coal beds under investigation are at depths ranging from 500 to over 1000 m, on the dip side of some existing mines which will not be economically viable for exploitation of coal in the foreseeable future, and the cut-off grade of methane in the coal beds is 5 cubic meters per tonne of coal with potential recovery of 20-50 per cent.

(c) *High sulphur coal*: During combustion of sulphur-bearing coal, the resultant sulphurous smoke causes air pollution as well as corrosion in the boilers. The sulphur also causes acidity in the mine waters. Sulphur is found in coal in three forms—pyrite, sulphate and organic sulphur. Sulphur content is believed to be high in coals deposited at shallow depths in neutral to weakly alkaline environment. Of the three forms of sulphur, organic sulphur poses problems for utilization particularly when its content exceeds 2 per cent. The pyrite and sulphate can be reduced by nascent hydrogen such as through treatment of finely crushed coal in acid solution containing granulated zinc and chromium powder. Experiments on removal of pyritic sulphur and sulphate sulphur through bacterial leaching by means of 'Thiobacillus Ferro-oxidants' and certain micro-organisms present in coal, have been reported to be successful. Possibility of recovering some sulphur as byproduct can also not be ruled out. But attempts to remove organic sulphur has not met with any significant success.

Chemists at Southern Illinois University, USA, reported in 1990 a new desulphurization method (Meyers-Read process), which used sulphonate compounds as reagents. These were mixed with water slurry containing finely pulverized high-sulphur coal, and air was circulated though the mixture causing cleaner coal to rise to the surface which could be collected and pelletized. Total sulphur content was claimed to be reduced to 50 per cent. But further developments regarding its commercial viability is not known.

2. Mining and Handling Rejects

During mining and handling, considerable quantities of coal dust are generated. These can be utilized by various technologies. Also, possibility of making coke out of non-coking coal dust has not been ruled out—at least in laboratory scale. Further, during mining of coal, in many mines, methane gas is emitted which is not only considered as waste but also highly hazardous. Besides, post-mining left over coal and mine sludge also constitute wastes, because considerable coal is lost for ever in them.

(a) *Formed coke*: It is an unconventional fuel prepared by mixing coal or char fines and then pelletizing or briquetting by using some binder like pitch or tar. On carbonization, the product can serve as a substitute of coke. But the technology is not cost-effective enough to be used in bulk quantities in conventional blast furnaces. However, it may be useful charge in manufacture of low-bulk, high value products like spheroid grade iron.

(b) *Small briquettes*: The process of briquetting consists in applying pressure to a mass of coal particles—with or without addition of a binder—to form a compact agglomerate with thermal value. Small briquettes for domestic purpose are produced in round or ovoid shape. This shape requires the mould to be like two cups, and it has the advantage that the pressed briquette falls freely from the underside of the mould.

(c) *Stamp-charged coke*: In stamp-charging technology, coke is made using inferior quality friable type of coking coal blended with prime coking coal up to the extent of about 25 per cent. The mixture is compacted into a coal cake by stamping method and then side-pushed into the coke oven (cf. in case of prime coking coal, the charge can be thrown directly from the top). The side-pushing facilitates use of the coal cake which otherwise would have crumbled, if top charged.

(*d*) *Coke from non-coking coal powder*: Some experiments were conducted in U.S.A in the past. It involved hydrogenation under very high pressure (200 atmospheres) with a catalyst. It was found to be too costly to be of any practical utility.

(*e*) *Coal dust injection*: Now-a-days, non-coking coal dust is used as a substitute for coke in some blast furnaces for pig iron manufacturing. This is known as 'coal dust injection (or CDI) ' or 'pulverized coal injection (or PCI)' technology. According to the World Coal Institute statistics (April, 2004), the quantity of PCI coals used in blast furnaces in the world has increased from 10.5 million tonnes in 1990 to 25.7 million tonnes in 2001, Japan, South Korea, U.S.A., France, Germany and Italy being the leading consuming countries. One tonne of PCI coal replaces 1.4 tonnes of coking coal.

(*f*) *Coal mine methane*: Theoretically, coal mine methane or CMM can be regarded as a subset of coal bed methane or CBM, being the name given to CBM that is released due to mining activities. In practice, however, CMM has a distinct identity of its own. Unlike CBM, it occurs in coal seams with limited permeability and is a hazardous waste generated automatically in many mines—irrespective of whether one wants it or not. In U.S.A., during 2003, approximately 1.1 billion cubic meters of CMM was produced.

Recovery technologies under research, are directed at:

(*i*) vertical or horizontal drilling deployed in advance of mining, injection of water at high pressure into the coal seams to fracture the seam and then pumping out the water to enable the methane to flow into the well.

(*ii*) drilling vertical or horizontal wells and introducing steerable motors to recover methane during mining.

Compared to CBM, quality of CMM is lower. Nevertheless, it is useful for a number of applications like power generation, heating, coal drying, boiler fuelling and industrial processes.

Considerable research and development works have been carried out in Germany on utilization of CMM. Three options have been explored:

(*i*) generation in boilers, of either heat or electricity;

(*ii*) combined heat and electricity production in prime movers (gas engines);

(*iii*) production of secondary fuels by conversion of CMM to methanol or its upgradation towards natural gas.

Out of these, as in 2003, the 3rd route was the most important one, and the 1st one is the second most important. The 2nd option has not been found to be economically viable.

(*g*) *Recovery from left over coal*: During exploitation of cut-and-fill stopes in underground mines, coal pillars are left out in order to prevent subsidence of surface land after the mines are worked out and abandoned. Substantial quantities of coal are lost for ever in these pillars. In Babnizu mine, Iran, a new approach has been adopted resulting in 50% recovery of the coal without any danger to the land above. The technology involves controlled blasting, followed by sealing with brick wall and steel.

In Central Mining Research Institute, India, bacterial technology for recovering left over coal from abandoned coal mines has been investigated with encouraging results. Methanogen bacteria has been found to be effective.

(*h*) *Biogasification of mine sludge*: According to research conducted in Central Mining Research Institute, India, mine sludge from coal mines can be harnessed for economic use through biogasification with the help of the microbe "Sporotrichum Purverulesstum".

3. Washery Rejects

These comprise high ash coal—more often coking coal in India—generally belonging to size range 3-25 mm. Both the middlings and the tailings constitute the so called wastes. Researches to utilize them have been directed mainly along the following two lines.

(*a*) *Oleoflotation*: Central Fuel Research of India (CFRI), in early 1980s, developed what is called oleoflotation, which is a modified version of froth flotation. It is essentially an oil agglomeration technique. Here, middlings are first ground to fine size. The fine coal particles are then flocculated by adding certain types of oil and centrifuged with coarser coal. Flocculation brings the coal particles together and separates them from the dirt. The extremely fine clay matter can be removed by this technique. But large scale industrial application of this technique has not been reported.

(*b*) *Fluidized bed combustion (FBC)*: When an evenly distributed flow of air or gas is passed at low velocity through a bed of fine particles of sand, the particles remain still. As the air velocity is increased slowly, the particles are first lifted, and then gradually a stage is reached when the entire mass of the particles is churned in a suspended state. It then appears to behave like a fluid in turbulent motion or a boiling liquid. When fuel is added to this bubbling bed, it gets distributed uniformly and if the bed is hot enough, combustion can be sustained. This is, in essence, the principle underlying FBC.

In an FBC system, the bed consists of an inert substance like sand or ash. Coal is crushed to 10 mm. size. A mixture of hot flue gas and air is blown through the system. The velocity of the gas is maintained at an optimum level. Too high a velocity will carry away the particles and too low a velocity will not fluidize the bed. Furthermore, if the coal contains a high proportion of fines, the bed velocities must be reduced to avoid excessive elutriation of unburnt carbon, and this may necessitate use of finer particles in the inert bed. The bed design calls for a compromise among particle size, pressure drop and fluidizing velocity. The bed temperature is maintained at 850° C to prevent clinkering of the ash. The low temperature pollution also reduces NO_X pollution. A fluidized bed has the following unique properties:

- as it behaves like a liquid, the bed level can be controlled;
- the solid particles move around very quickly so that good mixing between the gas and particles is achieved;
- because of this rapid movement, the bed temperature is uniform and easily controllable;
- heat is transferred rapidly to objects immersed in the bed.

The major advantage due to the above properties is significant increase in the burning efficiency of the coal. And this increased burning efficiency compensates for the low thermal value of the high ash inferior coal, thus opening up the possibility of utilization of the washery rejects for power generation. Coal containing as high as 70% ash and as low as 1700 kcal/kg thermal value can be effectively burnt. This technique was originally developed in UK in the late 1960s, and now it is being used in UK, USA, China, etc. In India its use is limited, and 10-15 MW power plants based on this technique are in operation.

It has been claimed that even high-sulphur coal can be used in FBC systems. In such cases, limestone or dolomite is added to the bed in order to fix up the sulphur by way of formation of calcium sulphate. This however, increases the overall cost of the process.

(c) *Back filling*: Fine coal processing wastes (FCPW) containing 65-70% solids are regularly used for back filling of underground room-and-pillar panels in Illinois mines in USA.

(d) *Bacterial demineralization*: Central Mining Research Institute, India has carried out investigations in this field. A bacterial species "Pseudomonas" has been identified for demineralizing coal washery rejects.

(e) *Reject recycling*: In India, washery rejects containing 55% ash are actually recycled to jig shells, and by jigging (see also the discussion on coal beneficiation earlier in this chapter), three types of products are generated. The products are:

   (i) fine coal fraction (0-6 mm) containing up to 48% ash,

   (ii) coarse coal fraction (6-15 mm) containing up to 45% ash, and

   (iii) the final rejects containing up to 68% ash which are made into lumpy forms by thickening with flocculents and filter-pressing, and then sold for use in brick kilns and as domestic fuel.

The first and the second types are both used by thermal plants located near the washery plants,

4. Wastes from Industrial Usage of Coal

Various types of waste are generated during use of coal in coke ovens to make coke, during burning of coal and during power generation. Efforts to develop technologies for utilization of these wastes are discussed as follows:

(a) *Coke breeze*: This term signifies the finely divided coke particles that are generated within coke ovens, during breaking of the oversized coke pieces and also during their handling. The coke breeze has high thermal value and can be used with advantage in lieu of coal where the charge has to be ground to fine size, such as in cement manufacturing, in boilers, in sintering, as a reducing agent in electric smelting, in chemicals manufacturing, and in foundry coke. Also by adding coke breeze to the raw coal mix charge in coke ovens, the quality of coke can be improved with reduction in its manufacturing cost. In electric smelting furnaces, coke breeze is mixed with larger sized coke to produce the right reactivity (on account of fineness) and right sensitivity (on account of air gaps).

(b) *Carbon Dioxide* (CO_2): CO_2 emitted by burning of coal is a major source of air pollution, and is the single most responsible agent for global warming or (as it is called now-a-days) the 'green house effect'. There are 5 green-house gases (or GHG), the emission of which is believed to be responsible for the green-house effect. These gases are: CO_2, NO_2, methane, CFC (chloro-fluoro-carbon), and water vapour. Out of these, CO_2 is the most widely produced gas directly related to burning of coal. It absorbs infrared rays coming from the sun, but does not allow it to go back to space; and an increased density of CO_2 in air causes increase in the temperature of the earth. It has been estimated that in the northern hemisphere, its density has increased by as much as 25% from 280 ppm to 350 ppm during the last 200 years or so. It has also been estimated that while the total emission of CO_2 in the world was 1640 million tonnes during 1950, it reached 5555 million tonnes in 1986 (out of which the contribution of fire wood alone may be 800-1600 million tonnes).

This CO_2 is now-a-days being regarded as an economic commodity. Its potential use is in food processing, fish farms, agricultural greenhouses, conversion to fuels, manufacture of stable products such as carbonate minerals and secondary recovery of petroleum (the gas industry routinely separates CO_2 from natural gas and it is then transported to market by pipeline). Scientists are therefore trying to develop what is called 'carbon sequestration' technologies to capture CO_2 from industrial emission streams and to store it for future use. Methods currently used for CO_2 separation include:

- Physical and chemical solvents particularly monoethaloamine (MEA).
- Various types of membranes.
- Absorption on to zeolites and other solids.
- Cryogenic separation.

However, application of these technologies for separating out CO_2 from high volume low CO_2-concentration flue gases is beset with problems of very high capital costs of installing the huge post-combustion separation systems needed, which are being addressed.

Regarding storage of captured CO_2, the following options are being considered.

- *Ocean storage*: This involves two main options. First is the dispersal of CO_2 as droplets at immediate water depths of around 500-1000 m; and the second is disposal at abyssal depths (5000 m or more) as liquid CO_2, but it may result in a measurable drop in the pH of seawater in the immediate vicinity of the injection site and impact on marine organism. Moreover, ocean is an open system and it would be difficult, if not impossible, to monitor the distribution and residence time of the stored carbon.
- *Mineral storage*: Mineral sequestration, also referred to as mineral carbonation, is the process whereby CO_2 is reacted with naturally occurring substances to create a product chemically equivalent to naturally occurring carbonate minerals. This is based on mineral feedstock, such as magnesium silicate (e.g., peridotites or serpentinites). The feedstock is mined, crushed and cleaned, if necessary, and then activated via a chemical or thermal treatment. It is then reacted in an aqueous CO_2 solution under high pressure to produce carbonate mineral plus sand in the form of finely precipitated solids, which

are separated from the liquid and sent to the final disposal area. Laboratory tests have shown that mineral carbonates can be formed rapidly, in time periods of about one hour. However, this requires intensive thermal pretreatment of the feedstock and subsequent carbonation at elevated pressures and temperatures. This would be cost-intensive. Also, the huge amount of materials handling that would be necessary to transport the silicates and dispose the carbonate would pose another major challenge to be resolved.

- *Geological storage*: By far the greatest potential for geological storage of CO_2 involves injection of compressed CO_2 into the subsurface, down to a depth of 600-800 m. An obvious site for geological storage is depleted oil and gas reservoirs. CO_2 would be compressed to a dense super-critical state. Some of it may react with the bedrock to form carbonate minerals, some would go into solution and remain stored for very long periods of time and can be monitored. Storing large amounts of CO_2 in deep saline water-saturated reservoir rocks, particularly sandstones, with the CO_2 stored as a result of hydrodynamic trapping, also offers great potential. One project is underway in the Norwegian North Sea Basin saline aquifer at a depth of around 1000 m below the sea floor. A comprehensive regional analysis of the storage potential of saline reservoirs has also been undertaken in Australia. Possibilities also exist for injecting CO_2 for enhanced oil and CBM recoveries.

Carbon sequestration for tackling CO_2 emitted due to industrial burning of coal is one of the most promising options. The International Energy Agency (IEA) has, in 2003, estimated that CO_2 sequestration systems could cost between $15 and $40 per tonne of CO_2 saved; transport costs were estimated at $1-$3 per tonne of CO_2 for each 100 km from the industry to the sequestration well; the injection costs could be $1-$2 per tonne.

(c) *Waste flue gas from power plants*: In Germany and Scandinavian countries, the sulphurous flue gases have been causing acid rains which in their turn have been damaging flora and fauna. A technology has been developed in those countries to produce synthetic gypsum from the waste flue gases. This gypsum is called "desulphogypsum". Subsequently, Japan and Austria have also adopted the technology. In this process, chlorine is first removed from the flue gas. Then it is sprayed by a mist of finely comminuted limestone slurry to form an aqueous $CaSO_3$ sludge. This sludge is then oxidized by passing air to yield gypsum. Desulphogypsum however is very fine grained and the plaster boards made out of this tend to shrink on drying.

(*d*) *Fly ash*: Ash is generated when coal is burnt in static beds (unlike in fluidized beds). The prefix "fly" is used for the lighter fraction of the ash, because during firing of coal in such beds, the velocity of the gases is great enough to lift these particles of ash from the bed, and 80-85% of the ash present in the coal is discharged to the atmosphere. In contrast, the ash left in situ after burning of coal is called 'furnace bottom ash (or FBA)" and the particles are relatively heavy. To prevent atmospheric pollution, the fly ash is collected with the help of special devices like electrostatic precipitator or ESP. Large volumes of fly ash are generated in thermal power plants where huge quantities of powdered coal are burnt round the clock for raising steam continuously, and this poses formidable problems of disposal.

Coal ash is a store house of trace elements. There is a great variation in the trace element concentrations in coals from different basins. But most coals contain very small, yet measurable quantities of metals and nonmetals—including rare earth elements, which occur within the crystal lattices of minerals associated with coal. Elements occurring in this manner are known as trace elements, and those elements in coal find their way into the coal ash. The trace elements in coal might have originated due to:

(*i*) their absorption from soil by the coal-forming plants,

(*ii*) their association with the mineral matter brought into the basin during the initial process of coalification,

(*iii*) their contribution by the surface and underground circulating waters, and

(*iv*) their deposition through the hydrothermal solutions during igneous activity.

The history of study of trace elements in coal and ash is not very old, having been started only in 1927 by Ramge. But the studies have so far revealed a wide variation in the concentration of different trace elements in samples drawn from different coalfields throughout the world. From Indian coals also, a number of trace elements have been reported. The following table shows the range of concentration of different trace elements reported from across the world and also their occurrence in Indian coals.

Trace element	Maximum value reported	Place where reported	Place of reporting in India
Antimony	0.3%	Dickebank seams of Ruhr basin in Germany	—
Argon	Trace	Mine gases from Ruhr in Germany	—
Arsenic	1.0%	Coal ash from Ruhr in Germany	—
Barium	5.0%	Coal ash from Ruhr in Germany	Umaria, Wardha valley, Damodar valley, Son-Mahanadi, Rajmahal, Meghalaya, Upper Assam, and J & K coalfields
Beryllium	0.4%	Coal ash from Ruhr in Germany	Ghugus colliery of Wardha valley
Bismuth	0.2%	Coal ash from Ruhr in Germany	—
Boron	0.3%	Coal ash from Ruhr in Germany	Damodar-Koel valley, Son-Mahanadi valley, Rajmahal belt, Satpura valley, Wardha valley, Godavari valley, East Bokaro and Nichahama (J & K) coalfields
Bromine	4.1 ppm	Coal from Saratov area in Russia	—
Cadmium	65 ppm	Coal from Illinois basin in USA	—

Trace element	Maximum value reported	Place where reported	Place of reporting in India
Cesium	360 ppm	Coal ash from Czechoslovakia	—
Cerium	360 ppm	Coal from Pernik basin in Bulgaria	—
Chlorine	0.77%	Coal from Zwickau-Oelsnitz district in Germany	—
Chromium	11.3%	Coal ash from Katharina seam of Ruhr basin in Germany	Umaria, Korba, Damodar valley, Son-Mahanadi valley, Rajmahal, Satpura valley, Wardha valley, Godavari valley, Meghalaya, Upper Assam, J & K coalfields
Copper	1.0%	Coal ash from Ruhr in Germany	Same as above
Dysprosium	39 ppm	Balkanbas basin in Bulgaria	—
Erbium	28 ppm	Coal ash from Plevno in Bulgaria	—
Europium	14 ppm	Coal ash from Plevno in Bulgaria	—
Fluorine	177 ppm	American coal	—
Gadolinium	81 ppm	Coal ash from Plevno in Bulgaria	—
Gallium	0.3%	Coal ash from Ruhr in Germany	Umaria, East Bokaro, Damodar valley, Son-Mahanadi valley, Satpura, Rajmahal, Wardha valley, Godavari valley, Meghalaya, J & K coalfields
Germanium	3.79%	Coal ash from Pirin deposit in Bulgaria	Talchir, Damodar valley, Son valley, Satpura valley, Wardha valley, Rajmahal, Meghalaya, Upper Assam coalfields
Gold	0.194 ppm	Unspecified	—
Hafnium	2.2 ppm	Coal from Appalachaean coalfield in USA	—
Helium	4.0%	Mine gases from Ruhr basin in Germany	—
Holmium	31 ppm	Coal ash from Plevno in Bulgaria	—
Indium	2.0 ppm	Coal ash (place unspecified)	—
Krypton	Trace	Mine gases from Ruhr in Germany	—
Lanthanum	220 ppm	West Bokaro coalfield in India	Damodar valley, Son valley, Satpura valley, Wardha valley, Rajmahal, Godavari valley, West Bokaro, Upper Assam, J & K coalfields

Lead	3.1%	Coal ash from Ruhr in Germany	Damodar valley, Son valley, Godavari valley, Mahanadi valley, Satpura valley, Wardha valley, Meghalaya, Upper Assam, J & K coalfields
Lithium	0.05%	Coal ash from Karvina region in Czechoslovakia	—
Lutetium	4.8 ppm	Coal from Balkanbas in Bulgaria	—
Manganese	2.2%	Coal ash from Ruhr in Germany	Damodar valley, Son valley, Godavari valley, Mahanadi valley, Satpura valley, Wardha valley, Meghalaya, Upper Assam, J & K coalfields
Mercury	12 ppm	Coal from Illinois in USA	—
Molybdenum	0.1%	Coal from Ruhr in Germany	East Bokaro, Raniganj, Jharia, Rajmahal, Son valley, Meghalaya, J & K coalfields
Neon	1.2%	Mine gases from Ruhr basin in Germany	—
Neodymium	156 ppm	Coal ash from Plevno in Bulgaria	—
Nickel	1.6%	Coal from Ruhr in Germany	Umaria, Korba, East Bokaro, Damodar-Koel valley, Son-Mahanadi valley,Satpura valley, Wardha valley, Godavari valley, Rajmahal, Meghalaya, Upper Assam, J & K coalfields
Niobium	0.1%	Coal ash from Brazina in Czechoslovakia	East Bokaro, Damodar-Koel valley, Son-Mahanadi valley, Satpura valley, Wardha valley, Godavari valley, Rajmahal, Upper Assam, J & K coalfields
Palladium	0.2 ppm	Coal ashes	—
Phosphorus	0.74%	Coal ashes from Karvina in Czechoslovakia	—
Platinum	0.5 ppm	Coal ashes	—
Praseodymium	32 ppm	Coal ash from Plevno in Bulgaria	—
Radium	5.6 ppb	Coal from Meszka in Poland	—
Rhenium	328 ppb	Central Asian coal	—

Rhodium	0.2 ppm	Coal ashes	—
Rubidium	111 ppm	Coal from Pirin in Bulgaria	—
Samarium	19 ppm	Coal from Pernik in Bulgaria	—
Scandium	0.3%	Coal ashes	—
Selenium	8.1 ppm	Coal from Appalachaean coalfield in USA	—
Silver	10 ppm	—	Korba coalfield
Strontium	0.1%	Coal from New South Wales in Australia	East Bokaro, Damodar-Koel, Rajmahal, Satpura, Son-Mahanadi, Wardha Godavari, Meghalaya, J & K, Upper Assam, Umaria
Terbium	4.4 ppm	Coal from Balkanbas in Bulgaria	—
Thallium	3 ppm	Coal from North Caucasus in Russia	—
Thorium	52 ppm	Coal from Pirin in Bulgaria	—
Tin	0.1%	Coal ash from Ruhr in Germany	East Bokaro, Damodar-Koel valley, Son valley, Rajmahal, Wardha valley, Godavari valley, J & K coalfields
Tungsten	0.43%	Coal ash from Bulgaria	—
Uranium	1.34%	Coal ash from Arizona & New Mexico in USA	—
Vanadium	8.64%	Coal ash from Russia	East Bokaro, Damodar-Koel, Rajmahal, Satpura, Son-Mahanadi, Wardha, Godavari, Meghalaya, J & K, Upper Assam, Umaria
Yttrium	800 ppm	German & British coals	East Bokaro, Damodar-Koel, Rajmahal, Satpura, Son-Mahanadi, Wardha, Godavari, J & K, Upper Assam, Umaria coalfields
Ytterbium	38 ppm	Coal ash from Plevno in Bulgaria	—
Zinc	0.7%	Coal ash from Ruhr in Germany	Damodar-Koel valley, J & K coalfields
Zirconium	0.7%	Coal ash from Katharina seam of Ruhr in Germany	East Bokaro, Damodar-Koel, Rajmahal, Satpura, Son-Mahanadi, Wardha, Godavari, Meghalaya, J & K, Upper Assam, Umaria coalfields

However, potentiality of so many trace elements notwithstanding, their recovery from fly ash has not been possible. Only Germanium has been reported to have been recovered from flue dust on a commercial scale in England. R & D activities and practical trials, nevertheless, point towards a few prospects of commercial utilization of fly ash as follows.

(i) *Insulation bricks*: In such bricks, the raw materials are required to be subjected to a very high temperature treatment for chemical transformation. Since fly ash has already undergone that treatment in the boiler, it may prove to be suitable for making such bricks. In India, Bharat Heavy Electricals Limited (BHEL), Tiruchirapalli is reported to have developed a process.

(ii) *Building bricks*: This is by far the most promising and most talked about area of fly ash utilization. Pioneering research was carried out in Central Fuel Research Institute (CFRI), Neyveli Lignite Corporation (NLC) and National Council of Cement & Building Materials (NCBM) in India. Fly ash bricks are being manufactured and used in China, Australia, India, etc. It has been estimated that 180 billion bricks are manufactured in India every year. Conventional bricks consume top soil, which is very precious for agriculture and forestry, and 200 tonnes of coal for every million bricks for burning. Fly ash can replace it partly (clay fly ash bricks), or fully. Further, fly ash contains some unburnt carbon, which may provide part of the heat needed (in case of fired clay fly ash bricks) for brick firing, thus saving on coal. Fly ash bricks have been found to match with, and may even be superior in quality to conventional bricks in terms with strength, water absorption, smoothness of surface, dimensional tolerance and economy. The policy of the Government as in 2004 is that an addition of a minimum of 25% fly ash in bricks, tiles, blocks within a radius of 50 km from coal and lignite based power plants is mandatory; and for this purpose, the power plants have to supply ash free of cost. A new technology called "Fal-G" using fly ash, lime and gypsum is being popularized. In this technology, the raw materials are ground and water is added to obtain a semi dry mass. The mass so obtained is shaped into bricks by machine moulding and then the pressed bricks are subjected to specific curing cycle in sun or in air and steam, to gain the required strength.

(iii) *Concrete products*: Fly ash can replace a part of the cement in mass concrete. Addition of fly ash improves some of the properties of concrete like compressive strength, finish, impermeability, etc.

(iv) *Portland cement*: Fly ash can be used as either partial or complete replacement of limestone clinker. It can be mixed with clinker and then ground. Certain percentage of ash in cement does not alter the properties and suitability of the latter. As per the status in India as in 2004, the usage is to the extent of 15-35% of the total raw materials.

(v) *Asphalt paving*: The fine size of fly ash may be of some advantage if it is mixed with bitumen. The voidage in the surface of road may decrease and the durability may increase.

(vi) *Sub-base for road making*: In some countries, fly ash is being used for this purpose. The availability of fly ash in the thermal power plants in huge bulk may prove to be of particular advantage for its use in India.

(vii) *Roads & embankments*: It has been reported that 50000 cubic meters and 285000 cubic meters of fly ash have been used in costruction of Okhla fly-over bridge at Delhi and LPG plant of Indian Oil Corporation near Badarpur respectively.

(*viii*) *Land reclamation*: Fly ash is used for filling up depressions in land surface. But the dust problem in dry dumps and fear of contamination of the water percolating through the filled up land have served as a deterrent to its large scale use for this purpose.

(*ix*) *Soil nutrients*: The trace elements of fly ash may correct some nutrient deficiencies of soil. It is successfully used as a source of essential plant nutrients like calcium, magnesium, potassium, phosphorus, copper, zinc, manganese, iron, boron, molybdenum and also for boosting crop growth and yields. Further, fly ash being alkaline, may be suitable as an additive to acidic soil. It has been successfully used to raise teak plants, cotton crop, various horticultural plant species and forestry in different places in India.

(*x*) *Fillers:* In this use, not fly ash itself, but one of its derivatives called 'cenosphere', may be suitable. Cenosphere is a silicate glass filled with nitrogen and CO_2, and it is produced due to conversion of a portion of the fly ash during the combustion process. The trapped nitrogen and CO_2 make cenosphere lighter than water, and this lightness combined with chemical inertness may prove to be of some advantage as a filler in plastic, rubber, adhesive, etc.

(*xi*) *Mine stowing*: Fly ash has been successfully used for this purpose in countries like Hungary, USA, Germany, Poland, etc. In India, a mixture of fly ash and sand in the ratio 50-60%:40-50% by weight has been used on trial basis as a substitute of sand in stowing in Singareni collieries.

(*xii*) *Source of iron*: In Romania, laboratory experiments were successfully conducted to blend up to 30% of ferrous fly ash from power station (46.71% Fe) with steel shop flue dust (64.71% Fe) and pellet making.

(*xiii*) *Synthetic zeolites*: Zeolites is a complex compound of aluminium and silicon having high cation exchange capacity (CEC) and high micro-sieving efficiency due to large pore volumes. Consequently, it is an excellent sorbent with ability to absorb transition metals. It can play a very important role in nuclear waste processing by removing lead. Now, fly ash contains alumino-silicate glass or mullite ($Al_6Si_2O_3$). It has been possible to synthesize low Si/Al ratio Na-rich zeolites from fly ash by either (*a*) treating ash with concentrated NaOH solution at elevated temperatures ranging from 150-200°C and at high pressure, or (*ii*) microwave radiation and fusion with NaOH followed by hydrothermal treatment.

(*xiv*) *Source of alumina*: It contains significant percentage of alumina, and has been experimented for its extraction.

(*xv*) *Paint* : It may be possible to use it in paints.

In India, coal contains high ash and hence the volume of fly ash generated in boilers is also very high. In India, the maximum annual rate of generation of ash is about 100 million tonnes. As in March, 2004, total accumulated ash has been estimated at 1500 million tonnes over 65000 acres of land in 85 utility thermal power stations, and its rate of utilization is not more than 4 per cent. The policy of the Government in vogue in 2004 is to encourage use of fly ash-based products through various concessions and other measures.

SUBSTITUTION

Factors

According to a United Nations Economic Commission for Europe (UNECE) report of 2002, coal is still essential to global economic and social progress. It accounts for 25% of commercial energy demand worldwide, with 38% of global electricity generated from coal. Coal is also a key requirement for two other building blocks of modern society—the production of steel, with 70% of total global steel production dependent on coal, and cement. But yet, there are some driving factors for substitution of coal as follows:

1. ***Thermal value***: The following table indicates the thermal values of different commodities used in everyday life, as standardized by the U.N.O. Coal itself has substituted some traditional fuels on account of its higher thermal value, and now on account of the same reason, coal is getting substituted by other commodities.

Type of fuel	Global average thermal value	Indian average thermal value
Petroleum products	10,440-11,135 kcal/kg	—
Natural gas	12,135 kcal/kg	8,000-9,480 kcal/cu.m (at 15°C, 13.25 m bar, dry)
Hard coal	7,000 kcal/kg	5,000 kcal/kg
Lignite (brown coal)	2,695-5,700 kcal/kg	2,310 kcal/kg
Fire wood (fuel wood)	2,331-3,600 kcal/kg	4,750 kcal/kg
Charcoal	—	6,900 kcal/kg
Electricity	860 kcal/kwh	—
Liquefied petroleum gas (LPG)	10,800 kcal/kg	—
High speed diesel (HSD)	10,200 kcal/kg	—
Kerosene	10,300 kcal/kg	—
Light diesel oil (LDO)	10,300 kcal/kg	—
Fuel/Furnace oil	9,800 kcal/kg	—
Naptha	10,500 kcal/kg	—
Petroleum coke	8,000 kcal/kg	—
Bagasse	3,800 kcal/kg	—
Waste paper	3,200 kcal/kg	—

Source: (*i*) Indian Petroleum & Natural Gas Statistics; 1990-91, Ministry of Petroleum & Natural Gas, Government of India.
(*ii*) Indian Cement Review, December, 2003

2. ***Chemical value*:** The utility of coal, because of its chemical derivatives, is virtually uncontested. Only the coal-based chemicals used in fertilizer manufacturing, have some noncoal alternatives.

3. ***Cost*:** In some areas, coal is not locally available, and cheaper easily available commodities may be used as substitutes of coal in spite of their inferior quality.

4. ***Conservation*:** In pursuance of national conservation policy, use of substitutes may be encouraged by the governments, even if substitutes may not be as effective as coal in their performance.

5. ***Pollution*:** This is by far the strongest complaint against use of coal. Systematic monitoring of the atmospheric concentration of carbon has alarmed the scientists in particular, and the people in general, about the contribution of coal to the global warming or the green-house effect (this has been elaborated under 'Carbon dioxide' in the sub-chapter 'Utilization of Wastes'). It has been estimated that since the Industrial Revolution till the end of the 20th century, the global average temperature has increased by 0.7°C due to man-made GHGs, and another 0.5°C increase may happen in the near future due to human activities during that period. The total and per capita carbon emission from fossil fuel during 1960 and 1987 are as given in the following table.

Country	Total carbon emission in million tonnes in 1960	Total carbon emission in million tonnes in 1987	Per capita carbon emission in million tonnes in 1960	Per capita carbon emission in million tonnes in 1987
Australia	24	65	2.33	4.00
Canada	52	110	2.89	4.24
China	215	594	0.33	0.56
England	161	156	3.05	2.73
France	75	95	1.64	1.70
Germany (Erstwhile Federal Republic of Germany)	149	182	2.68	2.98
India	33	151	0.08	0.19
Japan	64	251	0.69	2.12
Poland	55	128	1.86	3.38
Russia (Erstwhile USSR)	396	1033	1.85	3.68
USA	791	1224	4.38	5.03

Source: (*i*) The Economist Book of Vital World Statistics, 1991, Times Book.

(*ii*) State of the World, 1990, W.W. Norton & Co.

It is evident that the general trend of carbon emission was increasing till about 1990. In 1990, the 'Clean Air Act' was enacted in USA. Since the cost of stabilizing CO_2 emissions was estimated by different countries like USA, Canada, etc., was substantial, some of the governments even mooted the idea of levying of a 'carbon tax' on the carbon content of fossil fuels. Since 1990's, a system of tradable carbon credit is being debated. A carbon credit is a unit that measures a specific amount of GHG reduction. These credits are generally represented as a GHG reduction equivalent to a tonne of carbon dioxide or carbon or methane. A country able to achieve reduction in GHG emission will earn 'credits', which it can sell to a country emitting excess GHG. The bench mark for measuring reductions and excesses could be the targets set in Kyoto Protocol to which many countries are signatories. During the early 21st century, the World Bank has set up a division named 'Carbon Finance Business Division' and has created a 'Community Development Carbon Fund (CDCF)'. Following India's signing of Kyoto Protocol, the World Bank has issued, in 2004, a letter of intent for the purchase of 800,000 tons of carbon credits from the Indian fly ash brick industry @ US $ 5 per ton of CO_2 equivalent, in recognition of the fact that these bricks manufactured with Fal-G technology, do not consume any thermal energy and is thus an emission abating activity. This system of trade in carbon credits could serve as both an incentive for reducing and a disincentive for exceeding emissions. On the whole, since 1990s, there is an increased worldwide awareness and concern about pollution caused by coal and also systematic monitoring, resulting in a decreasing trend in CO_2 emission vis-à-vis GDP growth. But still pollution remains a dominant factor for substitution of coal.

Substitutes

The followings are the current and potential substitutes of coal in different uses.

1. ***Locomotive and engines***: Steam locomotives and steam engines were the earliest industrial uses of coal. However, these have long given way to diesel locomotives and engines because of superior thermal efficiency of diesel. More recently, many of the diesel locomotives have been replaced by electric locomotives. This substitution has been prompted by the consideration that electric locomotives do not cause any air pollution unlike both steam and diesel locomotives. But the electricity used in these locomotives is to a large extent generated in thermal power stations based on coal. So, in a sense, it is a full circle taking us back to coal.
2. ***Domestic heating***: In this use, cost, thermal value and pollution have been the driving factors behind substitution of coal.
   (*a*) *Firewood:* This was the oldest fuel used in domestic heating right since the prehistoric era—long before coal came to our life. In fact, when coal was discovered, it promptly substituted firewood in many parts of the world. But in some parts of India—particularly in the rural areas—either coal is not available or it is very costly and beyond the reach of the poor mass. In such areas, firewood is still used as a cheaper substitute of coal in spite of lower thermal value and greater pollution risk.
   (*b*) *Charcoal:* It is partially burnt wood and has higher thermal value than firewood. This was also a traditional fuel since long and continues to be so in areas where coal is not easily available, or (even if available) is too costly.
   (*c*) *Lignite:* Its thermal value is lower than that of coal. But it serves as a good substitute of coal in areas where it is locally available. In India, Tamil Nadu and Gujarat are two areas where there is no coal deposit, but where lignite is abundantly available. Now-a-days, a smokeless product based on lignite is marketed in India under the trade name "leco".

(*d*) *Kerosene*: This petroleum product has many advantages over coal. For example, it is not smoky and it has higher thermal value. Its main disadvantage vis-à-vis coal is that it is liquid and so it cannot be transported or stored as easily as coal. But, with domestic stoves replacing traditional ovens in many rural homes in India, kerosene has carved its place as a domestic fuel.

(*e*) *Liquefied petroleum gas (LPG)*: Like kerosene, it is also a petroleum product. Its thermal value is much higher than that of coal, and it has practically no smoke and hence no pollution effect. Its main disadvantages are that it is highly volatile, and can be liquefied only under very high pressure; hence its transportation and storage requires specially built containers and it has to be burnt in special burners, thus making it a very costly domestic fuel. Nevertheless, it has become popular amongst the affluent people, particularly in urban India.

(*f*) *Biogas*: Gas generated from biomass, i.e., biogas is a non-conventional renewable fuel which the governments in India and some other countries are trying to encourage as a measure of conservation of nonrenewable sources of fuel like coal and also firewood (which is nonrenewable in the short and medium run). Biogas is generated from biomass which include animal wastes like cattle dung, sheep dung, pig dung and human excreta or night soil, and the wastes from various food and vegetable items like sugar cane, rice, groundnut, coconut, oil seeds, cotton, etc. Biomass is converted to biogas by either of the following two processes:

(*i*) *Thermo-chemical:* In this process, pyrolysis results in gasification.

(*ii*) *Biochemical:* In this process, organic substances are broken down to CO, H_2 and acetates, and then converted to methane by methanogenic bacteria.

Thermal value of biogas may be up to about 80% of that of natural gas. Biogas generation from various animal wastes have been attempted in Kenya, Iran and Mexico. But the most common animal waste used is cow dung, and China is the pioneer country in this. In India, biogas generated from cow dung is popularly known as "Gobar gas", and a programme of setting up gobar gas plants was initiated in 1962 by the Khadi & Village Industries Commission (KVIC). However, for economic operation of such plants, a continuous and consistent inflow of cow dung as well as proper maintenance of the plants are necessary requisites. It has been estimated that on an average, one cow may yield 300 liters of gas per day.

So far as biogas based on human excreta is concerned, some experiments in Maharashtra and Tamil Nadu were conducted in the past. Gas generation from bagasse (waste from sugar cane) has been attempted by National Sugar Institute, Kanpur. Experiments have also been conducted for generating useful gas from city garbage.

(*g*) *Solar heat*: It has been estimated that the quantity of solar energy reaching the earth's outer layers is 0.17 million megawatts. As much as 30% of this is reflected back into space as light; 47% is absorbed by the atmosphere, land and ocean surfaces and converted to heat; most of the remaining 23% is used up in hydrologic cycle, i.e., evaporation-precipitation, ocean currents, wind generation, etc.; and a small fraction is used in supplying energy for photosynthesis. The huge amount of solar heat is a renewable energy and it can be used as a substitute of coal in domestic heating. But the main drawback is that solar radiation falling on a given area of a rooftop or yard has very small concentration. The technology for increasing the efficiency of utilization of solar heat is based on

(*i*) flat plate collector system

(*ii*) flat mirror system

(*iii*) paraboloid mirror system.

The flat plate system is used with a series of pipes placed in a black painted box for water-heating purpose. The black colour helps to prevent reflection of heat and enhance its absorption. This technology has been used in Israel to desalinate water by evaporating saline water and then condensing it.

The flat mirror system is used for concentrating solar radiation in solar cookers or solar ovens.

The paraboloid or boot-shaped mirror further intensifies the trapped heat and is used in making solar furnaces.

3. ***Industrial heating*:** In this use cost, thermal value and of course pollution are the driving factors behind substitution of coal.

   (*a*) *Furnace oil*: This is the heavy fraction of the distillates of crude petroleum and has much higher thermal value than coal. Though its cost is high, it is very effective in initial firing of furnaces because comparatively small quantities are required.

   (*b*) *Natural gas*: It also possesses a much higher thermal value than coal. It is used as a substitute for non-coking coal in sponge iron manufacturing. The main problem is to transport and store this gaseous fuel.

   (*c*) *Solar heat*: Use of natural solar heat is very common in drying of salt pans, fly-ash bricks and washed china clay. The rate of drying is very slow; but very low cost and nil pollution more than offset the slow rate. The main limitation, however, is that this technique cannot be used in wet and cloudy seasons.

   (*d*) *Solid wastes*: Solid wastes of scrapped rubber tyres, agricultural wastes, etc., are used in partial substitution of coal as fuel.

4. ***Reductant*:** The followings are the most common substitutes of coal or coke as reducing agent.

   (*a*) *Charcoal:* Charcoal is comparable to coal. However, it is forest-based and its widespread use will be detrimental to the cause of forest conservation.

   (*b*) *Natural gas:* Being a hydrocarbon, it is more effective than coal as a reducing agent. Moreover, in countries like India, where huge quantities of natural gas produced as a co-product of crude petroleum in many wells is burnt out for want of storage facility, increasing use of this commodity in lieu of coal should serve the dual purpose of: (*i*) scaling down the wastage of natural gas and (*ii*) conservation of coal. But natural gas suffers from the handicap that it is difficult to transport and store. Its effective use depends on extensive pipeline systems of transportation, and its use is by and large limited to selected places near natural gas fields. Recently however, investments are coming forth for exploration and development of natural gas fields as well as for laying pipelines.

(c) *Coal tar & coal dust:* Technologies are now available for using coal tar and coal dust as substitutes for expensive coking coal. In India, Bokaro Steel Plant, as reported in 2004, is planning to upgrade its facilities using these technologies.

5. ***Nitrogenous fertilizers***: The nitrogen and hydrogen contents of coal are recovered in the form of ammonia which is converted to ammonium sulphate—a nitrogenous fertilizer. The potential substitutes in this use of coal are as follows:

(a) *Biofertilizer:* After generation of biogas, the residue that is left in biogas plants, is a nitrogenous manure. The advantage of this manure vis-à-vis coal-based fertilizer is that its production is very cheap and environment-friendly, and also it is porous with high capacity to hold water.

(b) *Lignite:* In India, lignite-based fertilizer is being manufactured and this is cost-effective in areas like Tamil Nadu and Kachchh in southern and western India respectively, where coal is not available, but lignite is abundant.

(c) *Naptha:* The term naphtha applies to a petroleum distillate covering the end of gasoline and the beginning of the kerosene range, and it is a volatile hydrocarbon mixture. Though small quantities of naptha may be obtained from coal tar, it is mostly derived from petroleum by fractional distillation. The refinery streams going into products like gasoline (ranging from pentane, i.e., C_5H_{12} to dodecane, i.e., $C_{12}H_{26}$) and kerosene (ranging from $C_{10}H_{22}$ to $C_{14}H_{30}$) are grouped under naptha. The naptha can be a base for manufacturing nitrogenous fertilizer and in this, petroleum can substitute coal. Hydrogen content of naptha is more than that of coal, and in production of ammonium sulphate $(NH_4)_2SO_4$ fertilizer, naptha can contribute the requisite hydrogen.

6. ***Electricity generation***: One of the most important practical uses of coal is converting its thermal energy to electricity and then harnessing that electrical energy to serve various end purposes. This conversion is achieved with the help of water which serves to transform thermal energy of coal into kinetic energy of steam for driving turbines. Thus, this use of coal for generation of electricity can be considered as an indirect one. Now, with a view to conserving the finite resources of coal and also to reducing the rising cost of transportation of coal to remote areas, a movement has gained ground to harness various alternative sources of electricity. Some of these sources are truly nonconventional. In India, so much emphasis and encouragement are being given by the Government that there is a separate ministry dedicated to promotion of nonconventional energy. The currently known potential substitutes of coal in this use are:

(a) Nuclear fuel

(b) Lignite

(*c*) Geothermal energy

(*d*) Hydroelectricity

(*e*) Diesel

(*f*) Solar energy

(*g*) Wind

(*h*) Ocean energy (Tidal, Ocean thermal energy conversion or OTEC, Wave, Salt gradient)

(*i*) Biomass and agricultural waste

(*j*) Bacteria

(*k*) Industrial waste heat

Out of these, nuclear fuel, lignite, biomass and geothermal energy can be harnessed for generation of electricity indirectly through steam generation, while the other sources do not require this intermediate stage. Further, except nuclear fuel, lignite and diesel, all other sources can be considered as truly renewable. The substitutes are described as follows:

(*a*) *Nuclear fuel:* Bombarding the nucleus of some elements with a free neutron causes fission of the atoms of those elements. This fission releases enormous quanta of energy, which can serve the purpose of steam generation for the eventual production of electricity. The most common material used as the nuclear fuel is uranium ore. Uranium is a mixture of 99.3% of U_{238} and 0.7% U_{235}. It is the latter which is easily amenable to fission and hence some degree of enrichment of the natural element is necessary to increase the concentration of U_{235}. It has been worked out that one gramme of U_{235} can substitute approximately 3 tonnes of coal for generating electricity. The fission takes place in what are known as 'reactors'. Broadly, there are 3 types of reactors namely, light water reactor, heavy water reactor and fast breeder reactor.

In the fast breeder reactor, plutonium-239 (which is the altered product of U_{238}) can be used as the fuel, and thus it is also a substitute of coal. Indian scientists are exploring the scope of using thorium-232 (found in monazite) as the fuel in fast breeder reactors. On being hit by neutron, thorium-232 is capable of changing to thorium-233, which, through radioactive decay, can change into fissionable U_{233}. Thus, monazite can also be considered as a potential substitute of coal.

The principal advantage of substituting coal by nuclear fuel lies in the fact that use of the latter does not involve emission of any air pollutant. However, the main disadvantages of nuclear reactors are the high capital cost and the problems of disposal of the extremely hazardous radioactive wastes.

(*b*) *Lignite:* Lignite is used in the same way as coal. Thermal value of lignite is lower than that of coal, and so lignite is an inferior substitute of coal. But its use is cost-wise advantageous in areas where it is more easily available than coal.

(c) *Geothermal energy:* Geothermal energy is the natural heat of the earth's core, and the source of most geothermal energy is—directly or indirectly—molten rock or magma lying beneath the earth's crust. Four types of geothermal energy are known: hydrothermal (hot water or steam at moderate depths, i.e., from 100 m to 4,500 m), geo-pressed (hot water aquifers containing dissolved methane under high pressure at depths of 3 to 6 km), hot dry rock (abnormally hot geologic formations with little or no water), and magma (molten rock at temperatures of 700°C-1200°C).

At present, only hydrothermal resources are used on commercial scale. When underground water comes in contact with the magma, hot steam and water are produced. When this occurs in large quantities and within a few kilometers of the earth's surface, the steam and hot water can be tapped by drilling, and utilized to turn turbines for generating electricity.

A second method by which geothermal energy can be harnessed is to tap the heat energy of the natural geysers. A geothermal power plant in operation in California, USA, is based on the geysers north of San Francisco.

A possible third method to harness this energy can be by circulating water through holes drilled into hot dry rocks under the earth's surface. The heat of the rock can turn the water into steam which can then rise through a second hole for running turbines. Further, the efficiency of turning water into steam would be very high if the hot dry rock could be fractured by some device (say, nuclear explosion), in which case the cracks would allow the water to come in contact with a large surface area of the rock material.

Geothermal power plants are in operation in USA, Italy, Japan, Mexico, Iceland, New Zealand and Indonesia. Besides, considerable research has been carried out in this field in countries like USA, Russia, Kenya and Ethiopia. In India, an experimental 1 MW plant was first started in Pugga valley, Ladakh. Now-a-days, regular geothermal drilling is a part of exploration programmes of Government departments.

The geothermal energy potential is very large. Even the most accessible part is believed to exceed the current world annual consumption of primary energy which is about 400 exajoules (1 exajoule = 10^{18} joules).

(d) *Hydroelectricity:* In this case, the kinetic energy of falling water is harnessed directly to run turbines. Natural water-falls, artificial dams and tunnels (constructed in meandering courses of mountain rivers) can help to harness this energy. In the latter method, a tunnel is constructed to directly join two bends of a river—one at higher and the other at lower altitude, thus increasing the speed and energy of the naturally flowing water manifold, and this system is common in some parts of the Himalayas in India. In China, small dams are constructed in small rivers to set up mini-hydroelectric power generation units for serving local needs. In India, the potential from mini- and micro-hydel projects has been estimated to be 10,000 MW.

(e) *Diesel:* Diesel is neither cleaner nor cheaper than coal, and hence is not a good substitute of coal for electricity generation. The principle is the same as that in case of internal combustion engines (e.g., motors). Diesel electricity generators are used in a limited way to supplement demands in small localized areas.

(f) *Solar energy:* The theoretical potential of solar energy is immense. But the potential is limited by certain physical constraints like daily and seasonal variation, variation due to latitude, weather condition and the diffuse character of solar energy requiring large land areas for its large-scale generation. The following table showing estimations in 1994 and 1998 by the World Energy Council summarizes this.

Solar energy intercepted by the earth	5.5×10^6 exajoules per year
Solar energy reaching earth surface	2.7×10^6 exajoules per year
Solar energy reaching land surface	0.8×10^6 exajoules per year
Maximum world potential after considering the physical constraints, but ignoring technological or economic constrains	49,837 exajoules per year

1 exajoule = 10^{18} joules

Solar energy is pollution-free, noise-free and infinite. The heat energy from sun can be converted to electricity by Solar Thermal Electricity Conversion (STEC). This conversion is effected with the help of photo-voltaic cells (more commonly referred to as solar cells) which were invented in the 1950s. A solar cell essentially consists of layers composed of crystalline silicon doped with phosphorus, boron, etc. The key material is the silicon which is a costly semiconductor material having the peculiar property of producing electric current while absorbing sunlight. Attempts have been made to replace crystalline silicon by amorphous silicon and gallium arsenide.

The invention of semiconductor dates back to the beginning of 1960s, when scientists of Bell Laboratories in New Jersey, USA, observed that a silicon wafer could generate an electric current when struck with sunlight. Henry Kelly explained the phenomenon thus: "Energy in light is transferred in electrons in a semi-conductor material when a light photon collides with an atom in the material with enough energy to dislodge an electron from a fixed position and to enable it to move freely in the material". In the solar cell, when the semiconductor receives light, positive and negative charge carriers are released, and the electric field between the two differently doped areas of the semi-conductors separate the free charge carriers, that are then transmitted to consumers through metallic conductors.

One of the formidable problems with solar cells is high cost. However during the recent years, the cost is tending to come down. The cells installed in the space-station Skylab, launched in 1973, cost $300 per peak watt (peak watt is the quantity of electricity that a cell can produce from direct sunlight). In 1976, the cost came down to $45 per peak watt of power; in 1978, it was $7 per peak watt.

The second most formidable problem is the efficiency. Initially only 6-7% of the incident solar radiation could be converted into electricity. Constant research in the direction of increasing the number of layers in a cell, and also innovations in the field of new semi-conductor materials, have led to increase in the level of efficiency to 13-14% and even more.

The third problem is the low concentration of sunlight falling upon an area of the earth's surface. The energy emitted annually by the sun is 3.85×10^{23} kw of which the earth receives 1.8×10^{14} kw. The average intensity of solar radiation in India is 500-600 calories/cm^2/day and India receives annually 5×10^{15} kwh of solar energy. Even if 1% of this energy could be utilized, that would have met India's all energy needs. Many methods of concentrating solar radiation using mirrors and lenses have been tried, but these need clear sky and expensive tracking equipments. In England, experiments on developing a new technique have reportedly been successful. In this technique, flat-plate fluoroscent collectors were used to collect the ever-present green light (present even in diffused sunlight) from the sun rays, and to lengthen its wavelength to produce a light that could be accepted by solar cells for conversion into electricity. The key reportedly lay in the invention of a special dye based on an unusual metal that had the property of absorbing green light and changing it into a light to which solar cells are receptive.

A solar cell is of the size of a coin or less. Usually a panel of solar cells are used to generate electricity. Solar cells are indispensable in space crafts and satellites. They can also be used in various appliances like solar pump, solar lantern, etc., it has been estimated that in India, there is a potential to generate 570000 MW of solar energy annually.

The photovoltaic programme started in India in 1976. The commercial operation of STEC in the world started in 1984 in California's Mojowe desert, followed by Japan (1000 kw, 1981), Italy (1000 kw, 1981), France (2000 kw, 1983), Spain (1200 kw, 1983). In Japan, it is reported, 500000 new homes with photovoltaic cells on their roofs have been built as at the beginning of 2005. In India, STEC started in 1989, when a 50 kw plant was set up in Haryana. But even in 2004, higher capacity plants are in only planning stage.

(g) *Wind energy:* Winds develop when solar radiation reaches the earth's highly varied surface unevenly, creating temperature, density and pressure differences. Tropical regions have a net gain of heat due to solar radiation, whereas polar regions are subject to a net loss. This means that the earth's atmosphere has to circulate to transport heat from the tropics towards the poles. Rotation of the earth further contributes to the establishment of semi-permanent, planetary-scale circulation patterns in the atmosphere. Besides these forcing agents, other factors such as topographical features and local temperature gradients alter wind energy distribution.

Wind power was used for propulsion of boats in Egypt nearly 5,000 years ago. In China and Iran, wind energy was used in the 4th and 5th centuries through the design of vertical axis sail type wind mills. Electricity was generated using wind power in 1880, and in USA wind-operated water pumps worked during 1940's. World's first megawatt range Wind Energy Conversion System (WECS) was designed in USA in 1941 generating 1.5 MW output. This unit consisted of 2 blades of 24 meters diameter and a turbine mounted on a 35 meters steel tower. Subsequently about 45 meters tall wind mill generating 2 MW of electricity has been constructed in North Carolina, USA. The WECS converts the kinetic energy of the wind to mechanical rotary motion. The energy that can be obtained from any wind is proportional to the cube of the velocity of the wind. Therefore a slight change in wind velocity can make a substantial difference to energy of the wind. It has been observed that a minimum wind speed of about 3-4 m/sec is needed to work wind mills for generating electricity, while a speed exceeding 28 m/sec may engender mini-cyclone conditions with risk of damage to the components.

In India, the average wind speed is low. But in certain areas (coastal and hill areas) at certain times of the day and during certain seasons, the wind speed is considerably high. In Gujarat, Maharashtra and Tamil Nadu a number of wind mills have been set-up for supplementing the conventionally generated electricity.

The output depends on: (*i*) number of blades, (*ii*) tower height (*iii*) blade diameter and (*iv*) wind speed. While the first three can be designed, the last factor is totally unpredictable, and hence it is difficult to regulate the output generation. For example, in 1989, in the wind energy station in Okha, Gujarat, the capacity varied from 11 KW to 55 KW depending on wind speed. In the same year, in another station located in Mandvi (Gujarat), the capacity of different generators varied from 22 KW to 110 KW (tower height 22 m, blade diameter 6-9 m, Number of blades 2-3); and it was estimated that by increasing the tower height to 30 m and blade diameter to 12 m, the capacity could be increased to 250 KW.

The World Energy Council (WEC) in 1994, has estimated the global theoretical wind energy potential as 640 exajoules (1 exajoule = 10^{18} joules). In India, the wind energy potential been estimated to be between 30,000 and 50,000 MW. But in practice, the low capacity of individual generators and natural irregularities of wind speed do not make wind energy a viable substitute of coal in electricity generation; at best it can supplement coal.

(*h*) *Ocean energy:*

Tidal energy: Tides are formed due to the energy transferred to the oceans from the earth's rotation through gravity of the sun and moon. The World Energy Council (WEC) has assessed the global potential of tidal energy to be 22×10^{15} watt-hours per year or 79 exajoules per year (1 exajoule = 10^{18} joules). The idea of harnessing the energy of tides in India was first mooted in 1971. The sea-washed swamps of Sundarbans in West Bengal, the Gulf of Kutch and the Gulf of Cambay were identified as the potential areas. The methodology consisted in channeling and storing of the tides for electricity generation mainly in the estuaries where tides are most active. However, in 1975, the National Committee for Science & Technology estimated that India's potential tidal power would not exceed 1000 MW. In China, the first experimental tidal energy electric power station went into operation in May, 1981 in Zhejiang province in eastern China. Tidal power is generated in small quantities in France, Russia and UK.

OTEC: This is the acronym of 'Ocean Thermal Energy Conversion'. As the name suggests, the technique uses the thermal gradient in the ocean due to the difference in temperature between the warm waters on the surface and the ice-cold waters at several hundred meters depth, to generate electricity. The first unit named 'Mini-OTEC", was successfully operated in 1979 on a US Navy barge off the Kona coast of the Hawaii island. Its generation capacity was 50 KW.

There are two different methods of extracting the energy available on the ocean surface, namely:

- Closed cycle system
- Open cycle system

In the closed cycle system, a low-boiling-point fluid such as ammonia or propane is evaporated by the warm surface water in a boiler resulting in a vapour at high pressure. This vapour pressure is used to move a turbine. The spent vapour is condensed by cold sea water pumped from a depth of 300-1000 m and then recycled. The principle is the same as that of a thermal power except that low temperature is involved and instead of coal and water, are used warm sea water and some low-boiling-point fluid respectively.

In the open cycle system, the warm sea water is used as the active fluid and is flash evaporated under partial vacuum producing low pressure steam which can move a turbine. The spent steam can then be condensed by cold sea water pumped from depth, to produce desalinated potable water. Though recovery of potable water is an additional advantage, this system requires a much larger generator than the closed cycle system.

OTEC suffers from the following disadvantages:

- The thermal gradient becomes very low at latitudes beyond 15°, where the requisite difference in temperature (i.e., 17-20°C) is achievable only at uneconomic depths (beyond 300 m), and its economic potentiality is limited to tropical regions only.
- Sufficiently deep waters must be available near the shore so that transmission of electricity to the main land is economically viable.
- The generator system along with necessary infrastructure has to be established on offshore platforms or ships, and hence is costly.
- The system must be sheltered from cyclonic disturbances.
- Transmission of electricity has to be through submarine cables which are costly to lay.
- The material of the transmission lines must be corrosion resistant and at the same time must possess high electrical conductivity in order to minimize loss of power that is generated in this system in small quantities.
- Exploitation of this source of electricity in a very wide scale has the risk of lowering the average temperature of the oceans, and it has been estimated that if this temperature decreases by 2°C or more, then there is a possibility of the Arctic glaciers to advance.

From the point of view of depth of ocean and nearness to shore, the eastern coast of India is more favourable than the western coast. Off the eastern coast, sufficiently deep waters are available within 22-54 kms range, while in case of the western coast, deep waters are rather far away—beyond about 150 kms. Some locations off Tamil Nadu, Andaman & Nicobar islands and Laksha Dweep have been identified as promising. The potential of OTEC off Tamil Nadu has been estimated to be 10000 MW. In USA, intensive programmes have been drawn up for development of this non-conventional source of energy. It has been estimated that the total world potential of OTEC is about 100 million MW out of which up to 10 million MW may be practically usable.

The World Energy Council (WEC) has assessed the global potential as 2×10^{18} watt-hours per year or 7200 exajoules (i.e., 7200×10^{18} joules) per year.

Wave energy: This is the mechanical energy from wind retained by waves. The oscillating movement of sea waves can be transformed into electricity. The World Energy Council (WEC) has assessed its global potential as 18×10^{12} watt-hours or 65 exajoules (i.e., 65×10^{18} joules) per year. Norway is the pioneer in this field. In India, in the early 1990's, such a plant of 150 KW capacity was constructed 45 meters in front of the break water off Vizhingam fishing harbour in Kerala. Wave power may be economical if the plants are built as part of harbour construction. The concrete chambers built for housing the turbine-complex can act as harbour wall simultaneously generating electricity.

Salt gradient energy: This is the energy coming from salinity differences between fresh water discharges into oceans and ocean water. The World Coal Council (WEC) has assessed its global potential as 23000×10^{12} watt-hours per year or 83 exajoules (i.e., 83×10^{18} joules) per year. There is no report of any practical utilization of this potential.

(*i*) *Biomass & agricultural waste*: Biomass is an important energy source in developing countries. Recycling trash and garbage to usable energy can help solve the world's waste disposal problem and at the same time, fight the fuel crisis. In the commercially successful system, the classified trash is burnt directly in furnace to produce steam for moving turbines as in the case of thermal power. In a number of North American cities, garbage-based energy is being used. India's potential has been estimated to be between 17000 and 60000 MW.

In one system investigated in Japan, electric terminals of tin-oxide and platinum are used on opposite sides of a battery, to which were added some special pigment, vitamin-C, some weak liquid acid and a bacteria named 'Rhodospirillum Rubrum'. The chlorophyll in bacteria on the side of the tin-oxide terminal releases electrons by the process of photosynthesis. The electrons flow from the tin-oxide terminal to the platinum terminal through an outer circuit. The electrons are then conveyed through the special pigment to the chlorophyll in the bacteria on the side of the platinum terminal, which also release electrons. Success of this system may also help cleaning of household waste material on which the bacteria thrive.

More common however, is to gasify the biomass through pyrolysis at 200-600°C, and then generating steam with the help of the gas, which is similar to producer gas. The steam can then move turbines and produce electricity. In India, as in 2002, biomass-based gasification cum power plants have come up in the states of Madhya Pradesh, Uttar Pradesh, Maharashtra, Tamil Nadu, West Bengal, Gujarat, Karnataka, Uttaranchal, Andhra Pradesh, and Haryana.

In Western Australia, a project has been started in 2003, to plant pine trees in a 21000 hectares area, which will be cut down over the next 25 years to produce 90000 tonnes of laminated veneer annually, and the residue will result in generation of 160000 tonnes a year of agricultural waste. This waste biomass will be utilized to generate electricity for 24000 homes.

In India, the wastes generated after crushing sugarcane in the sugar industries of Maharashtra, are being planned for use for power generation. The potential in co-generation by all the sugar industrial is estimated as 1200 MW, but the actual generation till early 2005 has been a meagre 35 MW.

(j) *Bacteria*: Bacteria-based battery is in a preliminary research stage, and is under development in USA and Japan. The output is very small, but nevertheless it is being conceived as a potential supplementary source of electricity for household consumption.

(k) *Industrial waste heat*: It is now possible to capture waste heat from industrial smokestacks and turn it into electricity with additional benefits by way of cutting carbon emissions drastically and reducing the toxic pollution of the atmosphere. In this heat-scavenging system, propane vapour in place of steam is used to turn a turbine and drive an electricity generator. This allows it to be driven by low temperature waste heat. In case of steam to turn a generator, it must be pressurized and superheated to around 650°C (below 450°C, the process does not perform efficiently). But unlike water, propane's properties are much more suited to electricity generation at lower temperatures.

3
CHAPTER

PETROLEUM

Petroleum derives its name from Latin words 'petra' meaning rock and 'oleum' meaning oil. Formerly, it used to be called 'rock-oil', now the unrefined substance is referred to as 'mineral oil' or 'crude oil'. It has also been referred to by various other names such as maltha, asphalt, mineral pitch, bitumen, ozokerite, etc. It is a naturally occurring easily inflammable liquid, varying in colour from light yellow to green, brown or heavy black and having density between 0.79 and 0.95 gm/ml, under surface conditions, while viscosity is generally in the range 0.025 to 0.25 poise. Chemically, it contains principally hydro-carbons (50-90%, rarely up to 98%), minor quantities of oxygen, nitrogen and sulphur, and also traces of metals like sodium, magnesium, calcium, strontium, copper, silver, gold, aluminium, tin, lead, vanadium, chromium, manganese, iron, cobalt, nickel, platinum, uranium, boron, phosphorus and silicon in the ash left after burning crude oil. The hydro-carbons may be of three types—paraffins, cycloparaffins and aromatics.

Most of the oil fields have been found in sedimentary rocks—ranging in age from Cambrian to Plio-Pleistocene, though a small number of them also occur in igneous and metamorphic rocks. The origin of oil is still a matter of speculation (though the overwhelming view is in favour of organic origin—from both plant and animal remains), but it is generally believed that the present locations of oil pools are not where they originally formed and some degree of migration has invariably taken place.

HISTORY

Petroleum, in some form or other, has been known since antiquity. In the Old Testament it has been referred to as 'slime', which has later been translated as 'bitumen', and in the Tower of Babel, the walls of Babylon, and other constructions, this material was said to be used as a mortar. Ancient Egyptians used bitumen for construction of buildings and boats, as well as preservation of mummies. Herodotus, in 450 B.C., referred to oil seepages in Persia and Greece, and also to the pitch spring of the Ionic island of Zante. The Chinese are believed to have first drilled for oil in 221 B.C., and the Japanese and the Romans are reported to have used it for illuminating purpose around the beginning of Christian era. It is believed that the people during those times depended on accidental finding of natural seepages, from which the lighter fractions evaporated leaving behind the heavier bituminous or asphaltic residue for their use.

More recently, during the 16th and 17th centuries, asphalt was reportedly used in America for repairing of boats. Subsequently, during the 18th century and the early part of the 19th century, petroleum assumed an added importance on account of its suspected medicinal value. So far as some of the other countries are concerned, the recorded discoveries and production date back to 1765 (Myanmar), 1771 (Galicia), 1788 (Hungary), 1811 (Trinidad) and 1857 (Romania), while the early hand-dug oil-producing pits of Baku in Russia are known since the 17th century (in fact, such pits are reported to have yielded over 3000 tonnes of petroleum in 1840).

Meanwhile, the frequent finds of oil in the surface seepages and salt wells of USA led to some serious thinking in the minds of some. A sample from Pennsylvania was tested and two useful products, namely kerosene and lubricating stock, were separated. A small company was formed to drill specifically for oil. Col. E. L. Drake was entrusted with drilling in Oil Creek, Pennsylvania. This well, now known as 'Drake Well', struck oil on 28th August, 1859 at a depth of only 69.5 ft, and heralded the beginning of the modern petroleum industry. That year, 270 tonnes of petroleum were produced from this well. Many more wells were drilled, not only in USA, but also in other countries. By 1900, eleven countries reported a total production of 20 million tonnes from drilled wells. The world production rose phenomenally to 869 million tonnes in 1956, and to 3,149 million tonnes in 1990.

At about the same time the Drake Well was drilled, the first successful experiment with internal combustion engine (I.C. engine) was conducted in Paris by Etienne Lenoir, which was subsequently refined in 1878 by Nikolaus Otto in Germany. But those I.C. engines were using town gas supplied through pipes, as fuel, and so were not suitable for locomotive engines. The suçcess of Drake Well, however, opened the flood gate of possibilities of building I.C. locomotive engines using petroleum as fuel. Initially, in the 1870's, kerosene was experimented with. Then in the 1890's, vapour of heavy oil (which was earlier considered as an embarrassing waste by-product of kerosene) in a jet of compressed air, was found to be an effective fuel for burning within the locomotive engine. In 1892, Rudolf Diesel of Germany refined it further based on thermodynamic principles of minimizing heat loss. More improvements followed. Meanwhile in 1885, in Germany, a lighter distillate of petroleum was effectively made use of by Gottilab Daimler and by Carl Benz to make motor cycles and motor cars respectively. In the 20th century, petroleum has become an essential substance in vehicular transport, and there is now no looking back in the growth of petroleum industry.

In India, petroleum was discovered in Assam in the 19th century. Lt. Wilcox observed a seepage in the bed of Buri Dihing river at Supkong in Upper Assam for the first time in 1825. Subsequently, C. A. Bruce (1828), Major White (1837), S. Hannay (1837-38 and 1845) and H.B. Medlicott (1865) reported oil seepages at various places including some coal outcrops. In 1866, M/S McKillop Stewart & Co. drilled several holes by hand—one of which was 102 ft deep—on both sides of Buri Dihing. A well was also sunk to 195 ft depth. But results were not encouraging. However, others began drilling at Makum (27° 18′N & 95° 40′E), where oil was struck on 26th March, 1867 at a depth of 118 ft. More drilling followed, and in January, 1868, the daily yield from different wells varied up to about 3000 liters. The development of petroleum industry in Assam was mainly due to the initiative of Assam Railway & Trading Co., though some drilling work was also carried out by the Assam Oil Syndicate Ltd. The concessions held by those companies were taken over by the Assam Oil Company Ltd., which continued drilling and production in the Digboi oilfield, and which erected a refinery also. Meanwhile, oil seepages in Badarpur in the Surma valley of Cachar district were probed jointly by Badarpur Oil Syndicate and Burma Oil Co. Ltd. during 1915 onwards. In 1917, production commenced, but the field was finally abandoned in 1933 because the wells dried up after yielding a total of about 266 million liters of oil.

After independence in 1947, the Government of India laid special thrust on the development of oil industry in India. In 1956, Oil & Natural Gas Commission (converted to Oil & Natural Gas Corporation Ltd. or ONGC Ltd. in 1993) was set up, and in 1958, Assam Oil Co. Ltd. was nationalized resulting in the formation of Oil India Ltd. (OIL). More oilfields were discovered in Assam, Gujarat and also in the offshore structures like Bombay High. Production increased manifold. An idea of the growth in production of petroleum in India is given in the following table.

Year	Approximate production of petroleum
1900	3.42 million liters
1910	16.00 million liters
1920	50.37 million liters
1930	220.77 million liters
1940	Over 300.00 million liters
1950	259 thousand tonnes
1960	454 thousand tonnes
1970	6.8 million tonnes
1980	9.4 million tonnes
1990	33.3 million tonnes
2000-01	32.43 million tonnes
2002-03	33.04 million tonnes

Note: Production figures prior to 1947 do not include those of Myanmar and Pakistan.

CRITERIA OF USE

Chemical composition of petroleum is principal criterion determining its utility. However, a few physical characteristics also play an important role.

1. Chemical Composition and Characteristics

(*a*) *Hydrocarbons (HC)*: Though petroleum contains minor insignificant amounts of oxygen, nitrogen, sulphur and trace elements, by far the principal constituents are the hydrocarbons, i.e., compounds composed solely of carbon and hydrogen. The percentage of carbon may be up to around 12-13%, while that of hydrogen up to 86-87 per cent. The number of possible naturally occurring hydrocarbons is practically infinite. Three broad groups of hydrocarbons have been identified in petroleum. These are:

I. *Saturated*: (*i*) Paraffins (C_nH_{2n+2}) which are open chain HCs, e.g., methane (CH_4), ethane (C_2H_6), propane (C_3H_8), nonane (C_9H_{20}), etc.

(*ii*) Napthenes (C_nH_{2n}) which are closed chain HCs, e.g., Trimethylene (C_3H_6), tetramethylene (C_4H_8), cyclopentane or pentamethylene (C_5H_{10}), etc.

II. *Slightly unsaturated:* (*i*) Aromatic or Benzenoid (C_nH_{2n-6}) which are closed HCs, e.g. benzene (C_6H_6), toluene (C_7H_8), xylene (C_8H_{10}), methyl derivatives of benzene (pentamethyl benzene, i.e., $C_{11}H_{16}$), etc.

(*ii*) Naphthalene ($C_{10}H_8$) and its derivatives.

III. *Unsaturated:* (*i*) Olefines or ethylenes (C_nH_{2n}) which are open chain HCs with a structure different from that of napthenes, though both have the same general formula, e.g., propylene ($CH_3CH{:}CH_2$), ethyl ethylene ($C_2H_5CH{:}CH_2$), trimethyl ethylene [$(CH_3)_2\ C{:}CH \cdot CH_3$] etc.

(*ii*) Diolefines (C_nH_{2n-2}), e.g., divinyl ($CH_2{:}CHCH{:}CH_2$) etc.

(*iii*) Acetylenes or alkynes (C_nH_{2n-2}), e.g., acetylene HC=CH, propyne CH_3=CH etc.

As the names indicate, the saturated hydrocarbons are characterized by comparable chemical inactivity, while the unsaturated ones are vigorously attacked by sulphuric acid, chlorine, bromine, iodine, potassium permanganate, ozone and many other chemicals. The aromatics are slightly reactive and can be slowly dissolved by sulphuric and nitric acids. Olefins are isolated by what is known as "cracking" of saturated hydrocarbons, i.e., by decomposition of such hydrocarbons by heat.

Diolefines absorb atmospheric oxygen more or less readily and some of these hydrocarbons polymerize spontaneously. Some acetylenes also polymerize readily, particularly when heated, forming aromatic compounds. In certain cases strong sulphuric acid causes such polymerization.

The chemical reactivity and solubility of the hydrocarbon derivatives are very important because on this is founded the petrochemical industry. The products of petroleum industry comprising the direct derivatives of petroleum and those of petrochemical industry comprising various chemicals synthesized from petroleum gases, form the basis of use of petroleum.

(*b*) *Sulphur*: All oils contain some amounts of sulphur compounds. However, in some oils, like the Iranian oil, sulphur content is extremely small—of the order of 1% or even less; while in others like Mexican oil, Canadian oil, etc., it may be up to 5 per cent. Sulphur compounds are spread through the entire range of petroleum fractions, but are dominant in the heavier ones. Generally, sulphur content and asphalt content are directly proportional. It is believed that the former is readily susceptible to oxidation and thus to transformation into the latter.

There are three ways in which sulphur-containing compounds in petroleum could originate:

(*i*) decomposition of proteid matter in the source organic material;

(*ii*) secondary action of inorganic sulphates (like gypsum), on the putrefying organic mass;

(*iii*) metabolism of bacteria contemporaneous with the algae from which the petroleum is supposed to have originated.

Sulphur plays a somewhat ambiguous role in the utility of petroleum, inasmuch as on the one hand it is objectionable in petroleum products; and on the other hand, if recovered as a byproduct—as in Madras Oil Refinery, India—it adds to the economic value of the petroleum.

(*c*) *Oxygen*: Oxygen occurs in the form of naphthenic acids and other organic acids, mainly in the nonvolatile asphaltic matter. Its possible origin is attributed to oxidation of the relatively more reactive hydrocarbons and sulphur compounds in petroleum, and also to some oxy-compounds that might have been present in the original source material.

(*d*) *Nitrogen*: Nitrogen may occur in minute quantities in some oils in the form of basic compounds.

2. Thermal Value

The definition of thermal or calorific value has been dealt with in the chapter on coal. This value is expressed in units of kcals/kg or B.Th.U/lb. As per the global norm suggested by the UNO, the thermal value of average crude oil is 10,175 kcals/kg or 18,315 B.Th.U/lb. However, this value for some of the derivative fractions of petroleum (like jet or aviation fuel) may be as high as 11,790 kcals/kg (or 21,222 B.Th.U/lb).

3. Specific Gravity

Specific gravity of crude oil varies from about 0.771 (some US and Sumatran oils) to about 1.06 (Mexican oil). However, for the same crude oil, specific gravity may vary from one hydrocarbon component to another, and it is this variation which plays an important role in fractional distillation of different useful hydrocarbon derivatives, by which the latter are differentiated into lighter and heavier fractions. For example, the specific gravity of heavy fuel oil fraction may be over 0.95 while that of gasoline may be below 0.65. Specific gravity is particularly influenced by packing within the molecules.

4. Boiling Point

This varies progressively along the range of hydrocarbon components of petroleum, and it is another important criterion by which fractional distillation becomes possible. For example, amongst different paraffins, boiling point varies from as low as (–) 37° C for propane (C_3H_8) to as high as 370° C for pentatriacontane ($C_{30}H_{62}$); amongst the olefins, it varies from (–) 48° C (propylene) to 73° C (tetramethyl ethylene); in diolefines from (–) 5° C to 126° C; in napthenes from (–) 35° C to 172° C; and in aromatics from 80° C to 249° C. In general, the boiling point of individual hydrocarbons increases with the molecular weight.

5. Viscosity

Viscosity is that property of a liquid which is a measure of its internal resistance to motion and which is manifested by its resistance to flow. Viscosity changes inversely with temperature, and directly with specific gravity. Generally, viscosity of hydrocarbons tends to increase with decrease in hydrogen content. High viscosity facilitates adhesion of oil and durability of an oil film. Viscosity of petroleum differs from one fraction to another. Presence of larger paraffin molecules like waxes influences viscosity.

6. Characterization Factor

Different crude oils are often compared with the help of this factor, which takes into account both specific gravity and boiling point. The formula for this factor is:

$$K = [3 (T + 460)^{1/2}]/S$$

Where,

K is the characterization factor;

T is the average boiling point in °F at atmospheric pressure;

S is the specific gravity at 60° F.

7. Dielectric Strength

Dielectric strength is the voltage that an insulating material can withstand before breakdown. It is a measure of the electrical insulation, and is expressed in terms of specific resistance. In some heavy oil fractions, it may go up to 6 million megohms per c.c.

8. Flash Point

It is the temperature at which the vapours from oil ignite and break into flames.

COMMON USES

Crude petroleum can be used in three forms as under:

1. Natural petroleum
2. Petroleum products
3. Petrochemical products

The common examples are discussed as follows:

1. Natural Petroleum

Now-a-days petroleum is not used in crude and natural form. However, from the ancient times until the beginning of the modern petroleum industry in the latter half of the 19th century, it was mostly used in this form. The common uses were (*a*) in lamps for illumination, (*b*) medicine and (*c*) as mortar or binder in construction of buildings, boats, etc. Some rare oils like those of Borneo and Sumatra, however, contain as much as 40% of light low temperature boiling fraction and can be used directly as a motor fuel.

2. Petroleum Products

The electron configurations of carbon and hydrogen enable these elements to combine in hundreds of ways to form different compounds. Through the process of fractional distillation, various mixtures of such compounds are recovered as distinct and usable petroleum products. Common examples of such products are as follows:

(*a*) *Light distillates*

(*i*) *Petroleum gas:* It is mainly used as a domestic fuel. It is used either in gaseous form which can be supplied through pipe lines or in liquid form (liquefied petroleum gas or LPG) which is supplied in cylinders.

(*ii*) *Petroleum ethers:* It is mainly used as a source of pentane (used in photometry) and cymogene (used in freezing machines for the production of ice).

(*iii*) *Motor fuel or Mo-gas:* It is used as a fuel for spark ignition internal combustion (I.C.) engines using carburetors (e.g., motor cars).

(*iv*) *Naptha:* Used as a source of hydrogen in $(NH_4)_2SO_4$ fertilizers.

(b) *Middle distillates*

(i) *Kerosene:* It is also called paraffin oil. It is used as a smokeless fuel in wick lamps for lighting and heating purpose. It can also be used in low performance spark ignition engines (e.g., tractors) using special type of vapourizers instead of ordinary carburetors.

(ii) *Aviation turbine fuel (ATF):* It is used as fuel for aircraft turbines and jet engines.

(iii) *High speed diesel (HSD) oil:* It is also called gas oil because of its earlier use in gas-making, particularly to augment supplies of coal gas during periods of heavy demand. Now-a-days it is used in large internal combustion engines like trucks, buses, etc., which do not use carburetors, and instead employ compression ignition and fuel injection. These engines require high compression ratio and high thermal efficiency.

(iv) *Light diesel oil (LDO):* It is used in slow speed engines like those used in agriculture, concrete mixing and other similar industrial machineries.

(c) *Heavy ends*

(i) *Heavy fuel oil:* Used in marine propulsion and electric-power generation.

(ii) *Furnace fuel oil:* Used for firing of furnace, etc.

(iii) *Lubricants and lubricating oils:* These are non-fuel products used for lubricating engines, switches, transformers, etc.

(iv) *Bitumen:* It is a non-crystalline dark brown to black coloured solid or semisolid material obtained from petroleum as a residue after vacuum distillation.

(v) *Petroleum coke:* It is a solid byproduct of the thermal cracking of petroleum. It is mainly composed of carbon and has a lower ash content than coal coke. Its principal use is in the manufacture of electrodes.

(d) *Other special products*

(i) *White oil:* These are used in absorption of the heavier easily liquefiable components of a mixture of gases. These are used for preserving surgical instruments and making ointments, disinfectants, scents, hair oils, cosmetics, medicines, etc.

(ii) *White spirit:* It is used for dry cleaning and as a paint thinner. It is an intermediate product between kerosene and mo-gas.

(iii) *Paraffin wax:* It is a crystalline solid product used for making candles, glazing paper, preserving stones, in water-proof coating of matches, as an electrical insulator, in floor and boot polishes, etc.

(iv) *Mineral jelly:* It is a semi-liquid petroleum product used as a base for ointments, polishes, etc.

(v) *Petroleum pitch:* It is obtained from asphaltic oil and is used in road and pavement making, anticorrosive coating of iron, as an electric insulator in laying of cables, etc.

The principal end-uses of the different petroleum products can be summarized as follows:

(i) Illumination and domestic heating

(ii) Transportation
(iii) Liquefied petroleum gas (LPG)
(iv) Lubrication
(v) Dry cleaning
(vi) Candle and model making
(vii) Road and pavement making
(viii) Medicines and cosmetics
(ix) Preservative and anticorrosive agents
(x) Polishing and glazing
(xi) Paints
(xii) Electricity generation
(xiii) Industrial machineries, tractors and marine engines
(xiv) Electrode
(xv) Explosive
(xvi) Industrial heating fuel
(xvii) Grease
(xviii) Photometry
(xix) Refrigeration
(xx) Flotation reagents
(xxi) Carbon black

3. Petrochemical Products

In the petrochemical industry, aromatics and olefins obtained from cracking processes, are the usual starting point. The principal products are:

(i) Solvents
(ii) Synthetic detergents
(iii) Synthetic resins
(iv) Synthetic rubbers
(v) Man-made fibres
(vi) Chemical fertilizers
(vii) Pesticides
(viii) Perfumery
(ix) Edible fats
(x) Explosives
(xi) Radiator antifreeze

These uses are discussed as follows:

(i) *Solvents*: Solvents may be of two types—pure hydrocarbon solvents and chemical solvents. The former type can be considered as direct petroleum product obtained in course of normal refinery operations; examples of this type are benzene, toluene, xylene, etc. The solvents of the latter type are synthesized from the petroleum gases ethylene, propylene and butylenes, and are actually petrochemical products. The most common chemical solvents are alcohols (ethyl alcohol, isopropyl alcohol and secondary butyl alcohol), ketones (acetone, methyl isobutyl ketone and methyl ethyl ketone) and glycol ethers.

These solvents are useful in the manufacturing of various end-products like paints, varnishes, printing inks, polishes, pharmaceuticals, cosmetics, etc.

(*ii*) *Synthetic detergents*: Till 1930's, the raw materials of synthetic detergents used to be vegetable/animal oils and fats. Later on, some alkylates, which are petrochemical products, came to be used. The important groups of detergent alkylates are the alkyl aryl sulphonates, alkyl sulphates, alkyl sulphonates, etc. Paraffin is oxidized to produce synthetic fatty acids, which on catalytic hydrogenation under pressure yields fat alcohols. These fat alcohols are treated with sulphuric acid, and then various additives are mixed to finally produce a detergent. The significant properties of a synthetic detergent are wetting power, emulsifying capacity, dispersing and protective colloidal action, dirt-absorbing capacity and foaming power.

(*iii*) *Synthetic resins*: In common parlance these are often referred to as "plastics". But in the true sense, plastics include compounds based not only on petroleum, but also on wood, vegetable matter and animal matter. All plastics including petroleum-based synthetic resins consist of very large organic molecules (macromolecules) which are built-up by polymerization of smaller molecules. In trade circle, bulk solid polymers, usually supplied to fabricators in pelletized form, are called 'resins'. Some of the characteristic properties are low specific gravity, easy deformability, resistance to chemicals, nontoxicity, electrical insulation etc. Synthetic resins include common products like polyethylene (or polythene), polystyrene, acrylo-nitrite, polyvinyl chloride (PVC), etc., and special products like polyphenylene oxide, polybutylene terephthalate, polyethylene terephthalate, polyacetal, polycarbonate etc. These are used in numerous consumer products.

(*iv*) *Synthetic rubber*: Synthetic rubbers are produced by a process of polymerization of butadiene and styrene (an ethylene product), isoprene, etc., with some catalyst such as sodium. On polymerization, a latex is obtained which contains macromolecules having a filament-like structure. This latex is stabilized and coagulated by addition of acids and salts, then washed and then finally dried to yield synthetic rubber. Various substances are added to improve properties of this rubber. For example, sulphur and mercapto benzothiazole are added under pressure at approximately 150°C temperature to vulcanize the rubber, as a result of which filamentary molecules become interlinked by sulphur molecules and the strength of the rubber increases. Synthetic rubber is extensively used for making motor tyres, conveyor belts, etc.

(*v*) *Man-made fibres*: Man-made fibres are subdivided into two main groups—fibres made from cellulose and the synthetic fibres.

Cellulose fibres or wood pulp is prepared by chemical processing of pine, fir, sugarcane waste, straw, maize, sunflower stalks, cloth pieces and similar materials containing cellulose. The chemical processing involves treatment with calcium bisulphate solution or a mixture of caustic soda, sodium sulphide, sodium sulphate, sodium carbonate.

This processing results in liberation of soft cellulose by decomposition of the bonding substance, i.e., lignin. Fibres can be manufactured by treating the cellulose with caustic soda (viscose process), copper sulphate and ammonia (cuprammonium process), or acetic acid, acetate anhydride and sulphuric acid (acetate process), as a result of which its molecules are rearranged. Cell wool and rayon are examples of the products made out of cellulose fibre. Of these, the product obtained by acetate process—the acetate rayon—have some relation with petroleum inasmuch as the processing chemicals acetic acid and acetic anhydride are petrochemical products.

The synthetic fibres (or synthetic polymers), on the other hand, are true petrochemical products. In this case the molecules are first synthetically built-up from the elements carbon, hydrogen, nitrogen and oxygen and then formed into macromolecules by polymerization. The hydrocarbons of some petroleum products like cyclohexanol are chemically synthesized with nitrogen- and oxygen-containing substances like ammonia, air, etc. There are four main groups of synthetic fibres: polyvinyl, polyamide, polyacrylic and polyester fibres. The earliest invented synthetic fibres namely nylon (1932) and perlon (1938) belong to the polyamide group, while terylene belongs to the polyester group. Polyacrylic fibres are based on intermediate product acrylonitrite which is produced from the petroleum product propylene. The common polyester fibre called polyethylene terephthalate is a product of polymerization of DMT (Dimethyl terephthalate) which in its turn is manufactured from the petroleum product ethylene.

(*vi*) *Chemical fertilizers*: In this use, the contribution of petrochemical industry is in the production of nitrogenous fertilizers namely ammonium sulphate, calcium ammonium nitrate, ammonium nitrate and urea. The basis of these products is ammonia (NH_3) produced synthetically from nitrogen (of air) and hydrogen of some petroleum product (e.g., naptha).

(*vii*) *Pesticides*: Some of the fungicides and solid fumigants are derived entirely from petroleum, while many others are petroleum-based intermediaries.

(*viii*) *Perfumery*: Here, the solvency power of special petroleum spirits is made use of to extract perfume of flowers.

(*ix*) *Edible fats*: Certain fatty acids such as palmitic acid ($C_{16}H_{36}O_2$), when combined with glycerin, yield edible fats. Researches have in the past been carried out to produce such acids by oxidizing those hydrocarbons of allied structure and molecular weight which are paraffin wax.

(*x*) *Explosives*: Trinitrotoluene or TNT can be prepared from the aromatic hydrocarbons derived from petroleum. Toluene is a common member of the aromatic group. The aromatics possess a low degree of chemical reactivity and are converted into TNT by sulphuric and nitric acids acting in conjunction on toluene. The relatively high proportion of carbon compared to hydrogen in the aromatics contribute to the smoky flame of TNT.

(*xi*) *Radiator antifreeze*: Ethylene glycol, which is a petrochemical product is used as a radiator antifreeze (radiator is an engine-cooling apparatus in motor cars).

SPECIFICATIONS OF USE

1. Natural Petroleum

In all probability, the uses were determined by trial and error, and there was no conscious specification of the quality. The liquid crude oil was used for illumination. In some cases, from seepages, the volatile components of oil were vapourized leaving behind the semisolid bitumen or asphalt, and this was used by ancient people as a binder and mortar.

2. Petroleum Products

Refining is the key to the derivation of and utilization of the various petroleum products. The process of refining may be by:

(*a*) physical means such as distillation, solvent extraction, etc., and/or

(*b*) chemical changes such as cracking, reforming, polymerization, alkylation, isomerization etc.

Distillation is the first operation in refining. Boiling point can be varied by adjusting the pressure, and this principle together with the boiling temperatures and specific gravities of different components of petroleum is made use of in taking off the different light and medium distillates leaving behind the heavy ends. If only the lightest fractions are to be removed, the process is called "topping" or "skinning".

Solvent extraction processes are used for quality improvement of distillates by physical removal of aromatic and sulphur compounds. Solvents like liquid dioxide at low temperature are used. In this process, the relative solubility of the undesirable components vis-à-vis the desirable ones, is made use of.

In 'cracking', molecules are broken down under high temperature (with or without a catalyst) into smaller units, and a new type of hydrocarbon namely olefin is produced. By cracking, light gases, petroleum coke, fuel oil, etc., can also be produced.

Reforming is a special type of cracking in which a heavy low-octane naptha is processed for octane improvement rather than volatility change (octane number is a measure of 'anti-knock' value of a motor fuel, i.e., the ability to resist the knock or sound produced due to its sudden and violent combustion in a spark ignition engine. For this measurement, a standard scale has been devised by assigning the value zero to heptane (C_7H_{16}) which has very poor knock resistance, and 100 to octane (C_8H_{18}) having a very high knock resistance. Octane number is the percentage of this isomer of octane in its mixture with heptane, the knock resistance of which matches with the test sample. For example, if the knock resistance of the test sample matches with that of a mixture containing 75% octane isomer and 25% heptane, then the octane number of the sample is 75).

The process of polymerization (spontaneous alteration of substances) was developed with a view to utilizing the light gases (olefins) produced from cracking and reforming. In this process light molecules of olefin combine over a catalyst to form a heavier liquid which can be used in I.C. engines. Alkylation is the coupling of an olefin and a butane (or isobutane) over a catalyst. Isomerization process is usually run in conjunction with alkylation to provide sufficient isobutane which has more reactivity than butane (isomerization is the process of producing a similar but new substance by rearrangement of atoms within the hydrocarbon molecules of the original substance). It is obvious that in all these chemical processes of refining, the chemical composition and reactivity play the most important role.

For producing marketable petroleum products, removal of sulphur is a must. Presence of even a small amount of sulphur compounds is strongly objectionable, because these have an offensive odour, and also on burning, the sulphurous gases may have corrosive effect on engines and equipments.

Apart from these general criteria, which are key to recovery of the various usable petroleum products, there are some desirable specifications for the products to suit their actual uses. These are discussed as follows.

(i) *Illumination and domestic heating*: Mainly kerosene is used for this purpose. The common use is in wick stoves and wick lamps in which the kerosene rises in the wicks due to capillary action. To facilitate capillary rise, viscosity of the fuel must be very low.

Another important specification is the burning characteristic. The fuel is required to burn with a bright steady flame devoid of smoke and with a low tendency to form char on the wicks. This is possible if the dominant constituent of the fuel is hydrogen, far in excess of carbon. The paraffin group of hydrocarbons satisfies this condition well, and kerosene is composed of a mixture of hydrocarbons belonging to this group—usually ranging from $C_{10}H_{22}$ to $C_{14}H_{30}$. Aromatic compounds are undesirable, because these contain relatively less hydrogen and more carbon.

Thermal value of kerosene is also important, because it is used primarily for heating and illuminating purpose. On an average, its thermal value is about 10,638 kcals/kg.

The very fact that kerosene is meant for ordinary household use, requires that it must not be highly inflammable, or in other words, it should have a high flash point of over 27°C (cf. its boiling point ranges between 150° C and 300° C). Usually, the commercial brands of kerosene have flash points much above this minimum—over 34°C.

(ii) *Transportation*: Petroleum products are used in (*a*) motor cycles, motor cars, etc., (*b*) large trucks, buses, locomotives, etc. and (*c*) jet engines. The specifications vary according to the type of engine. The first kind of transport use I.C. engines with carburetor, the second kind uses I.C. engines with fuel injection system (some later generation cars also use this system), and the third kind uses jet turbine engines also with fuel injection system. In all these engines, the thermal energy which is released when the fuel is burnt, is converted into mechanical energy.

(*a*) *Motor cars, etc.*: In the carburetor system of I.C. engine, the liquid fuel is distributed in the form of very tiny droplets in a stream of air inside a device called carburetor. These droplets quickly vaporize and thus a combustible mixture of fuel and air reaches the cylinder of the engine, it is compressed and finally ignited by an electric spark produced by a sparking plug—the air of the mixture providing the oxygen for combustion. This combustion produces gases which expand and move a piston that eventually moves the wheels.

The petroleum product used is mo-gas or gasoline. One of the most important specifications is that it must have a high octane number, i.e., high anti knock resistance, though actual knock rating depends on a number of factors like spark timing, load, air-fuel ratio, engine speed, etc., besides the octane number of the mo-gas. Sometimes mo-gas obtained as a primary petroleum product does not possess a sufficiently high octane number and various additives are mixed to improve this property. Tetra-ethyl lead (TEL) used to be the most common additive earlier. But since the lead was proved to be a pollutant, hazardous to health, and producing high-octane fuel is much expensive, many of the producers have taken to the easier way out by adding aromatics like benzene to increase the rating. The content of benzene in unleaded petroleum is 5 per cent. But again, benzene is also a carcinogenic substance as is lead. Efforts are therefore directed towards reducing the benzene content from 5 to 3 per cent. Till the early 1980s, oil refineries were producing fuel with octane number up to 80. Subsequently, the grade has improved to 87-93 range.

The compression ratio of the fuel-air mixture in the engine ranges from 4:1 to 10:1. the mo-gas fuel must be able to withstand this compression and must not explode too violently producing undesirable vibration and overheating, and at the same time the thermodynamic efficiency, i.e., the efficiency of the conversion of its thermal energy into mechanical energy should be as high as possible.

The ignition of the fuel in the cylinder is required to result in instant combustion. Hence, its flash point should be low (much lower than that of kerosene).

The droplets of the fuel in the carburetor must vapourize quickly. So its boiling point should be low. But it should not be too low to create problems in storage and transportation.

The thermal value of mo-gas should be high, because essentially it is the thermal energy that is converted into the mechanical energy required for efficient transportation of the vehicles. On an average, the thermal value is about 11,135 kcals/kg.

Too high a carbon content will not only result in smoke nuisance while burning, but also may be deposited on the walls of the engine system and reduce its efficiency. So, hydrocarbons belonging to the paraffin group (in which hydrogen is predominant compared to carbon) are the desirable constituents of mo-gas. Usually it contains a mixture of hydrocarbons ranging from C_5H_{12} to $C_{12}H_{26}$.

(*b*) *Trucks, buses, etc.*: In these, are used what is called diesel engine, which is also a kind of I.C. engine, but with a fuel injection device instead of a carburetor. Instead of a mixture of air and fuel, air alone is drawn into the cylinder and is subjected to high compression resulting in its heating to a temperature of 700-900 °C. Only then the fuel is injected into the cylinder and is broken up into droplets. Because of the high temperature, the fuel first vaporizes and then ignites spontaneously. Due to the high compression needed, the engine requires a heavy structure. But, on the other hand, there is minimum heat loss due to the spontaneous ignition and consequently high thermodynamic efficiency.

The fuel used in this type of engines is high speed diesel or HSD. Since the fuel is required to vaporize only under conditions of high temperature inside the cylinder (and not at normal temperature inside a carburetor), its volatility may be lower, i.e., its boiling point may be higher, than in case of mo-gas.

The compression ratio in HSD engines is very high ranging from 14:1 to 25:1, so the diesel oil must be able to withstand this compression without exploding suddenly and violently and yielding loud knocks. Also, since the fuel in the cylinder is, before ignition, in the form of liquid droplets, chances are high that after ignition there will be some unburnt droplets which may ignite subsequently with sudden violence causing knock. So, the HSD fuel must have high octane number, and for this purpose the primary product obtained from distillation of petroleum is usually subjected to reforming and treatment with various additives.

The main advantage of HSD lies in its relatively less cost of production compared to mo-gas. This is because, compared to mo-gas, much larger quantity of HSD can be recovered from a given quantity of petroleum. In Indian refineries, HSD constitutes on an average 36% of all the products taken together compared to only about 7% constituted by mo-gas. For the sake of economy, higher content of carbon than in mo-gas is tolerated. This causes somewhat less thermal value, and increased smoke. However, thermal value of HSD is reasonably high, being of the order of 10,700 kcals/kg.

The viscosity of HSD should nevertheless be not too high to cause difficulty in injection.

(c) *Aircraft*: A modern jet engine works on the same principle as in the case of HSD engines, except that the compression is effected by a turbine instead of a piston, and that the combustion of the fuel and resultant expansion of the gases are so sudden that gases go out at a very high velocity in the form of a jet. This high velocity produces the propulsive thrust.

The fuel used is aviation turbine fuel or ATF. For the high combustion rate required, it is desirable that the thermal value of the fuel is high. In ATF, the thermal value is of the order of 11,800 kcals/kg.

In high altitudes, the pressure is low and the fuel must not have a tendency to boil at the reduced pressure. At the same time, it must also not freeze under conditions of low pressure and temperature prevailing at high altitudes. A few minutes after take off, an aircraft reaches a level where temperature is below (–) 30°C. Therefore, a freezing point of below (–) 60°C is generally desirable. Besides, the octane number must be very high so as not to produce knocks and to withstand the high compression and very rapid combustion. This in turn requires that there should be an even mixture distributed throughout the cylinders. In other words, the vaporization of the different ingredients of the fuel should take place within a range narrower than in case of mo-gas.

ATF is seldom obtained as a primary product of petroleum refining, and considerable processing and treatment are necessary for imparting the right combination of specifications.

(iii) *Liquefied petroleum gas or LPG*: LPG is supplied in pressure cylinders and is a special fuel used for domestic cooking in gas-ovens. It burns without smoke and without plugging the oven with residual material.

For making LPG, moderately volatile fractions of petroleum containing only a little carbon are suitable. The moderate volatility, i.e., moderately low boiling point enables the gases to be liquefied under conditions of ordinary temperature, but high pressure, and also to be revaporized as soon as the pressure is released at the time of burning in the oven. Too high volatility (as in the case of methane) makes it difficult to liquefy, while too low volatility prevents easy revaporization. The low carbon (and relatively high hydrogen) enables the LPG to burn with a clean smokeless flame.

Methane (CH_4) and ethane (C_2H_6) are too volatile to be suitable for LPG manufacturing, while members of paraffin series of hydrocarbons higher than butane are too less volatile to suit the purpose. The principal constituents of LPG are therefore propane (C_3H_8) and butane (C_4H_{10}).

(iv) *Lubrication*: Lubricating oils may be of use in engines, machineries, electrical transformers, switch gears, etc. The specifications vary according to the operating conditions under which the lubricating oil is required to be applied. Lubricating oils are nonfuel, inert, viscous, heavy fraction of petroleum.

The most important specification is viscosity. Viscosity tends to change with temperature. So, lubricating oils are graded according to their suitability to different temperature ranges, and for each grade a high viscosity is to be maintained. Related to high viscosity is high specific gravity which thus becomes another important specification.

Within the specified range of temperature also, the viscosity of the oil should be resistant to change with fluctuations of temperature. It has been observed that paraffins fulfil this condition better than napthenes.

One of the functions of lubricating oils (particularly those applied to circulating systems like bearings, etc.) is to remove the heat from the system and keep it cool. For this purpose, the oil must not itself have any thermal value.

Another important criterion is the life of the lubricating oil. Under the influence of heat, air and combustion products, lubricants tend to undergo decomposition with consequent loss of viscosity and effectivity. High flash point is therefore a desirable specification. Additives have also been developed to retard this rate of decomposition and prolong the lives of lubricants with accompanied increase of cost. But in noncritical service, life of the lubricant may sometimes be compromised for the sake of economy, even at the risk of increases in the cost of repairs and replacements of machinery parts due to sub-optimum lubrication.

Wax is highly objectionable in lubricants. Wax has a low melting point, and hence a lubricant containing wax tends to congeal under low temperature conditions. At higher temperatures also, the molten wax, being itself low in viscosity, reduces the overall viscosity of the lubricant.

When a lubricant is meant for use in transformers and switchgears, it must, in addition to the other specifications, possess high dielectric strength so as to serve as an effective insulator as well. Oil with specific resistance as high 6 million megohms/cc have been employed in high voltage installations. But such high dielectric strength can only be achieved if the oil is absolutely free from water.

(v) *Dry cleaning*: For this purpose, the oil is required to possess high solvency power, so that the grease and dirt from leathers, clothes, etc., can be dissolved and removed. At the same time the oil must be colourless so as not to leave any stain on the washed articles. Also, the oil should have a low viscosity so that it spreads easily and evenly on the entire surface of the article being washed. Besides, the oil should have a fairly low volatility (of the order of 80-95 °C) so as to facilitate quick drying of the washed articles.

Refined petroleum distillate called 'white spirit' is used for this purpose.

(vi) *Candle and model making*: The petroleum product used in candle-making is the wax (or paraffin wax), which is a solid hydrocarbon obtained from the heavy distillates of crude oil. Paraffin wax may also be obtained after purification of a naturally occurring dark brown coloured mineral wax called 'ozokerite' or 'ceresin'. Low melting point, ready setting and colourlessness are the most important criteria.

Low melting point enables the candles to remain in solid form at ordinary temperature, and also, during burning of the candles, it enables the molten wax to lose viscosity quickly, thus facilitating its capillary-rise through the wicks. Ready setting property of the wax facilitates its casting into different shapes of models and candles. Finally, due to its colourlessness, it is possible to add various colours and to make models and candles of any desired colour.

(vii) *Road and pavement making*: High viscosity, high melting point, dark brown to black colour, inertness and low cost are the principal criteria. The heavy residue left after distillation of petroleum, which is variously called as 'petroleum pitch', 'asphalt', 'bitumen' and 'asphaltic bitumen', is the material used for this purpose. It is resistant to most chemicals (it is soluble only in CS_2).

(viii) *Medicines and cosmetics*: The petroleum product called 'white oil' is used for the medicinal and cosmetic value of its chemical ingredients, and another product namely 'mineral jelly' or 'vaseline' is used as a base for medicinal and cosmetic ointments for external application.

White oil is obtained from light lubricating oils by drastic refining. It is substantially colourless. Vaseline is a gel-like substance derived from heavy lubricating oils. It is semisolid, amorphous and colloidal at ordinary temperatures; if heated, it liquefies, but again re-jellifies on cooling. Further, its characteristic chemical neutrality, freedom from unpleasant odour or taste, and light colour make it highly suitable to its use as a base of ointments. Examples of medicines and cosmetics include nasal sprays, sun-tan lotions, disinfectants, hair oils, scents, etc.

(ix) *Preservative and anti-corrosive agent*: Such substances have to be chemically non-reactive and waterproof, so that they can protect the coated surfaces against atmospheric oxidation and actions of other acids, alkalis, etc., as well as keep them dry. A waterproof coating applied on fresh food stuff, can preserve its freshness by preventing its moisture from escaping.

Amongst the petroleum products, asphalt (or bitumen), white oil and wax find such applications. Bitumen is used mainly for protection of iron pipes, poles, etc.; white oil is used for preserving surgical instruments; and wax is used as a coating on card boards, stones and match sticks. Coating of wax is also applied to preserve the moisture and freshness of cheese, egg, etc. Both asphalt and wax can be used as a coating of electrical cables and articles, because both of them have good insulating property.

(*x*) *Polishing and glazing*: The main purpose is to impart smoothness and glaze to a surface. Among the petroleum products, varnish and wax possess this characteristic. Varnish is prepared by solvent treatment of asphalt, and is used for polishing wood. Wax is used for polishing very soft or delicate substances like paper, bottle labels, frescoes, etc.

(*xi*) *Paint*: White spirit is used as a thinning agent for paints. In this use, the low viscosity, white colour, chemical inertness and capacity to dissolve the paint oil are the main specified properties.

(*xii*) *Electricity generation*: Diesel generators are essentially fuel injection system I.C. engines directly coupled to a generator. The fuel used is high speed diesel (HSD) and its specifications are same as discussed earlier in case of similar engines employed in heavy vehicular transports like buses, trucks, etc.

(*xiii*) *Industrial machineries, tractors, marine engines etc.*: These employ fuel injection system I.C. engines, but these are slow speed engines. The compression ratio of these engines may be of the order of 4.5:1 or so-much lower than those used in cars, trucks, buses, etc. Consequently, the thermal value and octane number required for the fuel are less than in case of high speed I.C. engines. Light diesel oil (LDO), which is a petroleum distillate heavier and less volatile than HSD, serves the purpose of fuel for such engines. Only disadvantage is that at the time of start, the low compression ratio does not produce enough heat in cold cylinder to vapourize the droplets of the liquid LDO fuel. To overcome this problem, mo-gas is used at the start to generate sufficient heat in the cylinder. Thereafter, once the engine is warmed up and it starts running, it can switch over to LDO. Since such engines normally run continuously for long periods of time and are not required to frequently stop and start, LDO suits the purpose well, and by virtue of its less cost, effects economy.

(*xiv*) *Electrode*: In power intensive metallurgical process—such as aluminium smelting, very high quality carbon anodes are employed. The ash content in petroleum is negligible. The common carbon-containing substances like coal-coke or graphite contain considerable ash and no degree of processing can bring them down to the level of petroleum. Petroleum coke which is a solid product composed of almost pure carbon is preferred in this application.

(*xv*) *Explosive*: Explosives like nitroglycerine and geencotton mixture have a tendency to decompose. The decomposition products cause further decomposition and thus the rate of decomposition goes on accelerating. Mineral jelly, by virtue of its content of unsaturated compounds can absorb the traces of the decomposition products thus inhibiting the process of decomposition. Thus, its incorporation serves to stabilize the explosive mixture.

Iodine is a desirable constituent in petroleum jelly used for this purpose. As a result of the spontaneous decomposition of these explosive mixtures, nitrous gases are formed. These gases are absorbed by iodine, which thus helps to further stabilize the explosive mixture.

(*xvi*) *Industrial heating fuel*: Specifications of fuel for initial firing of furnaces, engines, etc., need not be as stringent as those for continuous running of them. On the other hand, economy and making best use of the distillates otherwise not usable, come to be of prime concern. It is in this context that fuel oil (or furnace oil), which is a heavy distillate of petroleum with low volatility, high viscosity and high carbon and ash contents, find application for this purpose.

Due to low volatility, the fuel oil needs longer time to vaporize and catch fire. But as an initial operation only for firing a furnace or an engine, this disadvantage is tolerated. The high viscosity by itself does not pose much of a problem, because once the oil is heated, viscosity is reduced sufficiently to enable the oil to be efficiently sprayed or pumped. The high carbon and ash are tolerated, because the firing operation does not last too long to create much of a smoke nuisance or to deposit much of carbon and ash on the walls of furnaces and engines.

(*xvii*) *Grease*: Greases are prepared by incorporating additives into viscous petroleum fractions. These may be grouped into three classes:

(*a*) admixture with solid lubricants such as graphite, sulphur, mica, talc, asbestos;

(*b*) mineral oils thickened with soaps of different bases;

(*c*) blends of residual stocks with pitches, waxes, fats, etc.

Viscosity and chemical inertness should be the most desirable properties of the petroleum product used in grease preparation.

(*xviii*) *Photometry*: This is a minor use. Pentane, separated from petroleum ethers, has been used for production of a standard reference flame of one candle power for the purpose of measuring the candle power of town gas.

(*xix*) *Refrigeration*: Another component of petroleum ether namely cymogene has a very low boiling point, and hence it can be quickly vapourized accompanied by quick loss of latent heat from the surroundings. It has been used to a small extent in freezing machines for the production of ice.

(*xx*) *Flotation reagent*: Xylene (C_8H_{10}) recovered from aromatic hydrocarbon derivative of crude petroleum, is a common reagent used in concentration of ores by flotation.

(*xxi*) *Carbon black*: Carbon black is a loose amorphous powdery and pure form of carbon used principally in rubber goods and also in printers' ink, pigments, etc. It is mixed with synthetic rubber tyres to impart abrasion resistance. Principal noncarbon components are oxygen (2.5%), hydrogen (0.5%) and sulphur. When petroleum is used as the feedstock for its manufacturing, it is also called 'oil black'.

For manufacturing carbon black, heavy fraction of petroleum with high carbon content, is suitable. Such oil along with air is fed into a reactor. Combustion of a part of the hydrocarbon raises the temperature to 1100-1700 °C causing decomposition of the unburnt portion of the hydrocarbon to carbon black. A water filter quickly cools the hot reaction products, and the finely divided 'black' is recovered. The yield may be of the order of 0.3-0.7 kg/liter of oil.

3. Petrochemical Products

As has been said earlier, the aromatics and olefins obtained from cracking processes are the usual feedstock for petrochemical products. Both these groups of hydrocarbons are more or less unsaturated and hence responsive to the action of various chemical compounds and elements. Essentially it is this property of chemical reactivity that is made use of in production of petrochemicals.

UTILIZATION OF WASTES

In broad sense wastes can be of three types—in situ waste, deleterious ingredients (mainly sulphur) and unconventional oil. All these types of wastes are now-a-days utilized to a large extent.

1. In Situ Wastes (Depleted Reservoir)

Underground petroleum reservoirs exist under high pressure acting from all sides. As soon as some opening is created for them through drilling of wells, the petroleum starts gushing out due to relief of the pressure. Thus, natural forces are the key to production of petroleum. However due to continuous depletion of a reservoir, a stage comes when the natural forces weaken and are no longer sufficient to push the oil up to the surface. Thus, the oil still left in the reservoir becomes useless for man and can be as an in situ waste, unless special techniques are employed for recovering the oil.

The enhanced oil recovery (EOR) methods employed to recover oil from reservoirs deprived of the natural pressure are secondary recovery techniques. Either gas or water is injected under high pressure through additional bore-wells so as to artificially augment the pressure in the reservoir, under the action of which recovery of oil improves. Currently, only about 35% of the oil in place is recovered by conventional production methods. Through the use of EOR methods this rate can be increased to as much as 65% of the original oil in place in a reservoir, although at higher costs of extraction.

2. Sulphur

So far as the deleterious ingredients in petroleum are concerned, sulphur is the most objectionable. It is found in petroleum in the form of different organic compounds such as thiophen, thioethers, mercaptan, etc., sulphur derivatives are removed by various oxidizing agents such as copper oxide, bleaching powder, sodium hypochlorite, potassium permanganate, etc. Sulphur is recovered as a useful byproduct from petroleum refineries.

3. Unconventional Oil

In addition to conventional oil, there are vast amounts of unconventional occurrences. These include oil shales, heavy crude oil, and natural bitumen (tar sands) containing extra heavy crude oil. These sources of oil are proving economic in favourable places, and further development may depend on higher oil prices, technological developments and long-term demand for liquid fuels.

(a) *Oil shale*: Strictly speaking, oil shale is not oil-saturated shale. Oil shale is a type of sedimentary rocks formed mostly under fresh or brackish water conditions, and they contain what is called 'kerogen'. Kerogen is a complex organic matter present in carbonaceous shales. It is insoluble in all common solvents, but on destructive distillation, yields oil, gas and acidic and basic compounds. Kerogen is formed by the biochemical and dynamochemical conversion of plant and animal debris. It consists of a mixture of carbon (77-83%), hydrogen (5-10%), oxygen (10-15%) and small percentages of nitrogen and sulphur, derived from a thoroughly pulped mixture of spores, algae, etc. It constitutes generally 10-35% of oil shale, the rest being clay and silt. On being heated strongly, oil shale yields a petroleum-like oil.

Oil shale beds are not ordinarily considered as an economic source of petroleum. However, in times of crisis, these can be mined by opencast or underground method (in much the same way as coal), crushed and heated in retorts. The products obtained are gases, crude shale oil and spent shale. Ammonia can be recovered as a valuable byproduct. The crude shale oil can be treated much the same way as petroleum to yield different refined products. The spent shale can be used to manufacture shale-lime bricks, hydraulic lime and cement.

During the 1990's, oil shale was produced in comparatively small quantities in China and Estonia. Estonia was the only country with an economy dominated by oil shale as a source of energy and has been the largest user of oil shale in power generation for more than 70 years. The annual production is of the order of 20 million tonnes.

(b) *Heavy crude oil*: Heavy oil is defined as high-viscosity crude oil with density less than 934 kg/m^3 (10-20° as per American Petroleum Institute or API standard). Genetically, heavy oil is formed by degradation processes from conventional oil resources occurring in shallow reservoirs. Currently some 8% of world oil production come from heavy oil reservoirs, with Venezuela, USA, Canada, Iraq, Mexico being major producers. Due to the nature of heavy oil, EOR methods such as steam flooding, hot water, polymer and CO_2 injection are generally required for its extraction.

(c) *Natural bitumen (tar sands)*: Tar sands are sands or sandstones impregnated with highly viscous extra heavy petroleum or semisolid bitumen with density greater than 1000 kg/m^3 (less than10° API). They are formed by thermal metamorphism and biodegradation of conventional oil deposits. Because of its nature of occurrence, it cannot be pumped through wells. Instead, such rocks can be mined by open cast or underground methods. Bituminous sandstone can be directly used as road metal, for paving and for roofing. Sometimes these can be crushed and heated to yield small quantities of oil. It has also been possible to separate the heavy oil from sand, clay and water by centrifuging, and to yield petroleum coke, lighter oils and sulphur. Natural bitumen typically contains high proportions of sulphur and trace elements including vanadium and nickel.

These have also been referred to as 'asphaltic rocks', and limestone may also similarly contain extra heavy oil.

The production of unconventional oil occurrences may adversely affect the local environment. Mining, conversion and upgrading can produce a range of toxic heavy metals and large quantities of solid and acidic liquid and gaseous wastes that need to be properly contained, cleaned and/or disposed of in an environmentally benign manner

SUBSTITUTION

Petroleum products have substituted many traditional substances since the nineteenth century, because of superior properties. However, a world-wide consciousness has begun arising about the limited resources of petroleum (that too concentrated in a limited number of countries, which thus enjoy enviable positions of economic and political power). Further, it is being increasingly realized that burning of petroleum contributes to environmental pollution in no mean a degree. These factors have prompted extensive as well as intensive researches to search for substitutes. Some of the substitutes being experimented with, are new and unconventional substances, while others are improved versions of traditional means. The trend of substitution of petroleum in its broad application areas are discussed as follows.

1. Transportation

(*a*) *Animals*: Horses, elephants, bullocks, camels, etc., are traditional means of transportation of men and materials. Even now, they serve as supplementary means mainly for relatively short distance transportation in rural areas, in forests, within factory premises, within farm lands, etc.—particularly in the less developed countries. According to a survey made in 1979, there were 80 million work animals in India—comprising 70 million bullocks, 8 million buffaloes, 1 million camels and 1 million horses—which together could provide 40 million H.P. energy (equivalent to 30000 MW electricity). According to the same survey, the number of animal-drawn carts in India was 15 million, and two-thirds of rural transportation of goods was by animal-drawn vehicles. Two surveys—one by Food & Agricultural Organization (F. A. O.) in 1953 and another by Punjab Agricultural University in 1970 gave the following comparative picture:

Name of unit	Average weight in kgs	Average H.P.
Bullock	400-700	0.75
Buffalo	500-900	0.75
Cow	400-600	0.45
Human	—	0.067
Tractor	—	22

In forests of Assam, elephants are deployed for transportation of timber. They were also used in a refractory plant in Rajgangpur, Orissa, for shunting of wagons.

(*b*) *Hydrogen:* Liquid hydrogen has been used as a fuel for launching rockets. Now, intensive research is going on in Japan, Germany, USA and other countries to make it a suitable fuel for cars. Principal advantage of hydrogen fuel over that of petroleum lies in its practically zero pollution effect, the products of its burning being only heat and water. On the other hand, the most formidable problems are its low flash point that renders it highly inflammable and highly risky to carry in tanks. Now, research is directed to development of special alloys which will be able to absorb hydrogen and then slowly release it for combustion. Another problem is concerning its usage in combustion chambers for airplanes—the steam produced may increase the risk of icing at high altitudes, thus increasing the chances of crash.

(c) *Ethyl alcohol*: Ethyl alcohol or C_2H_5OH is also called ethanol, and, more often, simply alcohol. It was the first organic chemical produced by man. It is volatile and combustible; it also has high octane rating. It is produced by fermentation of agricultural products like sugar cane, molasses, etc., which are rich in carbohydrates or starch (molasses is the final viscous mother liquor which is left after recovery of all the sugar from sugar cane juice and which is resistant to further crystallization). A mixture of ethanol and gasoline, called 'gasohol' is used as motor fuel in USA and Brazil. Although its opportunity cost tilts in favour of its use in food substance, it can serve as a partial substitute of gasoline or diesel in places where petroleum is either extremely costly or in short supply, and where molasses is available in plenty.

(d) *Methyl alcohol*: It is also called methanol, and it has the chemical composition CH_3OH. It is produced from inorganic carbon-containing sources like lignite, natural gas, syngas (mixture of CO and H). This can also be used as a motor fuel—either alone or mixed with ethanol, gasoline or diesel.

(e) *Natural gas*: Compressed natural gas (CNG) is now-a-days being used in a big way for running buses and other public transports—particularly in some of the highly polluted cities like Delhi, with a view to reducing air pollution due to vehicular emission. It has high octane number, and emission of CO and SO_2 is negligible. A train based on CNG has reportedly been running in Peru and another based on liquified natural gas (LNG) has been experimentally run in Delhi.

(f) *Electricity*: In railway transportation, electric locomotives have become more popular than diesel ones in India, because of:

- (i) the electricity can be generated from cheaper sources like coal, hydro-power etc.
- (ii) the former has zero pollution effect, and
- (iii) the former has greater load-bearing capacity. In road transportation also, battery-operated buses are used for plying over short distances.

(g) *Wind*: In maritime transportation, wind is the second oldest form of energy after muscle power. The traditional sailing ships have now given way to diesel-driven ships. It has been estimated that shipping accounts for 3-5% of global oil consumption, and that 30-40% of a ship's fuel costs should be saved if additional system to harness wind power were installed. Two systems have been mooted:

- (i) replacing traditional masts and sails with huge rotor blades encased in tube-like structures for harnessing wind power directly, and
- (ii) converting wind into energy with the help of turbine engines. By 1992, Japan and Germany have already developed a few wind-assisted vessels. At present, there appears to be a commercial future for wind power in yachts, trawlers and other small vessels.

(h) *Soyabean oil*: It has been experimentally used to run a modified diesel truck in USA. Although the products based on vegetable oil could be a renewable source besides being more eco-friendly than petroleum, the biggest hurdle is its unreliability at both high and low temperatures, and also the high cost of deriving vegetable-based motor oil. A lot of serious research is in progress in USA to overcome these problems.

(*i*) *Jatropha oil*: A potential bio-fuel, Jatropha Curcas is a plant that grows well on dry land. It has a 50-year life, and it yields up to five tonnes per hectare of oilseeds that can produce two tonnes of bio-diesel. Its use is free from CO_2 emission. Its byproducts can be used to make various products like soap, candle, glycerine and compost. An indigenously designed plant with 250 liters per day capacity is reported to be in operation in India in 2004.

(*j*) *Oil from plastic*: A new technology to produce oil from plastic waste has been developed in 2003, and a 5000-liter-per-day capacity plant has been commissioned in Butibori near Nagpur, India in April, 2005. Claimed to be first of its kind in the world, the technology, as reported, consists in shredding waste plastic articles of every-day use (*e.g.*, carry bags, PVC, bottles etc.), and then heating in absence of oxygen and in presence of a catalyst. The products comprise 80% liquid hydrocarbon, 15% LPG and 5% coke. One tonne of plastic waste may yield 1000 liters of oil, which can be an input for refinery.

(*k*) *Karanj oil*: In early 2005, a 7.5 kv capacity generatory based on oil extracted from Karanj seeds (a non-edible vegetable bio-fuel) has become functional in a remote village, Dharamgota, in Yavatmal district, Maharashtra, India. It has been claimed that 4 kgs of the seed can yield one liter of bio-diesel and 3 kgs of byproduct karanj cake that can be used as a fertilizer and pesticide.

2. Heating

Natural gas and petroleum can be used interchangeably depending on availability and cost. The thermal values of both are comparable, being of the order of 10,000-12,000 kcals/kg.

3. Petrochemicals

(*a*) *Natural gas*: The paraffins ethane (C_2H_6) and propane (C_3H_8) are common to both petroleum and natural gas. So, in the production of the chemicals based on these two hydrocarbons, natural gas can substitute petroleum. Such chemicals include synthetic fibers (polyester fibers, acrylic fibers, etc.), synthetic rubber (styrene-butadene), synthetic resin (polythelene or PVC), acetylene, detergents, acetone, various solvents, etc.

(*b*) *Coal:* Certain petrochemical products can be derived through distillation of coal tar. For example, ammonia which is used to manufacture ammonium sulphate fertilizer, may be manufactured from either naptha (a petroleum derivative) or coal tar. Other such chemicals include benzene, toluene, xylene, phenol, etc., which are used in the manufacture of various end-products like insecticides, detergents, explosives, paints, varnishes, printing ink, etc.

4

CHAPTER

NATURAL GAS

Natural gas (older name "marsh gas") consists principally of the most volatile hydrocarbons belonging to the paraffin family. It usually occurs as gas cap in the highest parts of the oil bearing horizons in reservoirs. Often natural gas fields are encountered where there is no associated oil. The hydrocarbon constituents are so volatile that they can remain in gaseous form even under the high pressure conditions prevailing in underground reservoirs. However, there may be some constituents of relatively lower volatility which remain dissolved in oil underground, but which become gaseous as soon as they escape to the surface through wells. Also there may be vapours of some easily liquefiable gasoline constituents like butane (C_4H_{10}), pentane (C_5H_{12}) and hexane (C_6H_{14}) admixed with natural gas. Such natural gas is called "wet" gas in contrast to the "dry" gas which is devoid of these higher members of the paraffin series. Now-a-days, natural gas is regarded as a valuable economic commodity, and today's fuel of choice. It is flexible to use, environmentally friendly compared to other fossil fuels, relatively abundant, with supplies perceived to be relatively secure and reliable. Consequently, it is used in a variety of sectors and applications, and experiencing significant growth.

HISTORY

Natural gas was known in the historic times, but not as a useful substance. It used to be associated with some mysterious deity. Emanations of natural gas in many spots have been burning since time immemorial, and these "eternal" fires are worshipped by people. At Baku in Caucasus, the Temple of the Fire Worshippers was built in such a spot. Similarly, at Baba Gurgur in Iraq, such an eternal fire was known during the time of king Nebuchadnezzar. In India also, such continuously burning natural gas fires have been deified and temples built over them (e.g., Jwalamukhi in Himachal Pradesh, Ankleswar in Gujarat).

Even in the earliest days of the oil industry, the economic importance of natural gas was not appreciated. In India the role of natural gas as a pressurizing agent in the recovery of petroleum in Digboi oil field began to be recognized towards the beginning of 20th century. But, still, its economic potentiality remained largely untapped. The problems of its storage and transportation contributed, to a great extent, to its wastage by allowing it to be burnt out. In the neighbouring Myanmar, natural gas is known to have been transported through pipeline and used as a fuel in cement works in around 1935. But in India, until recently, the natural gas emanating through oil wells was regarded as more a liability than an asset. Only a fraction of it has been utilized locally in and around oil fields. In 1965, the Assam Gas Company Limited (AGCL) —a Government of Assam undertaking, was set-up to transport natural gas to power, fertilizer and petrochemical industries, to tea gardens and to domestic consumers in Assam area. However, the optimum utilization of this valuable natural resource has received a real thrust in India only in 1984, when the Gas Authority of India Ltd. (GAIL) was established with a view to constructing a network of pipelines for distribution of natural gas along with facilities for storage, so as to promote growth of gas-based industries throughout the country. The most important and ambitious task undertaken by the GAIL is the construction of the 1700 km long Hazira-Bareilly-Jagdishpur (HBJ in short) trunk pipeline connecting Hazira in Gujarat and Jagdishpur in U.P. With this development, the Oil and Natural Gas Corporation (ONGC) of India has now taken interest in exploration and exploitation of exclusive natural gas structures like those of Bombay offshore, Tripura, Andhra Pradesh and Rajasthan, where there is no associated oil known to be available. Now-a-days, even submarine pipe lines are contemplated to connect Indian industries with natural gas wells of Oman.

The production statistics for natural gas is available since 1960 and the following table gives an idea of the growth of its production in India.

Year	Natural gas (utilized)
1960	147 million cubic meters
1970	67 million cubic meters
1980	1,462 million cubic meters
1990	12.464 billion cubic meters
2000-01	27.86 billion cubic meters
2002-03	29.97 billion cubic meters

CRITERIA OF USE

1. Chemical Composition

The hydrocarbons constitute the combustible gases. The principal constituent is methane (CH_4) which constitutes on an average 85% of natural gas. This is followed by ethane or C_2H_6 (10%), propane or C_3H_8 (3%). The balance 2% may comprise butane (C_4H_{10}), pentane (C_5H_{12}), hexane (C_6H_{14}), heptane (C_7H_{16}) and octane (C_8H_{18}). Natural gas often consists of some incombustible or highly toxic gases like CO_2, N, O, SO_2, H_2S, etc. The sulphurous gases are believed to be of volcanic origin deep down in the earth's crust. However the percentages of different components may vary from sample to sample. For example, in some natural gas, methane content may be as high as 99% or even more. It was found that samples of natural gas from Canada and America may contain inert gases like helium, neon, argon, etc. In India, a sample from Gogha in Vadodara district, Gujarat was reported to contain 0.8% He. It is the hydrocarbons that are responsible for most of the economic value of natural gas (a more detailed account of the hydrocarbons has been given in the chapter on petroleum). Helium, when present, adds to the value considerably. However, the other constituents are regarded as either deleterious or of no consequence.

2. Thermal Value

Hydrogen and carbon contribute to the thermal value of natural gas. Its thermal value is more than that of other gaseous fuels. Under normal conditions of temperature and pressure, it generally ranges from 850-1400 BThU/cft (cf. 583 for coke oven gas, 573 for coal gas, 300 for blue gas, 150 for producer gas). The thermal value increases when natural gas is under higher pressure, and it is very high when it is in liquefied state, because of the fact that its density increases and under pressure more quantity of gas can occupy the same volume of space.

3. Gaseous Form

Under normal temperature and pressure, natural gas is in gaseous form. This is due to low volatility of the main constituents methane [B.P.(–)159°C], ethane [B.P. (–)89°C] and propane [B.P. (–)42°C]. The extremely low B.P. of methane which is the principal constituent of natural gas, enables it to remain in gaseous form even under high pressure.

COMMON USES

1. Chemical Products

Various chemical products based on methane, ethane and propane are manufactured with natural gas as the starting feedstock. The important products are as follows:

(a) *Synthetic fibres* : Ethane on cracking ("cracking' has been explained in the chapter on petroleum) loses a hydrogen atom and yields ethylene (C_2H_5), which being unsaturated reacts readily to form first ethylene oxide, then ethylene glycol. Finally, the ethylene glycol is treated with di-methyl terephthalate or DMT, and polymerized to yield polyester fibre (or poly-ethelène terephthalate or PET). Similarly, propane is cracked to yield propylene from which first acrylonitrile, and then acrylic fibre is produced.

(b) *Synthetic resins:* Ethylene can be converted to ethylene di-chloride. From this, vinyl chloride is produced, which on polymerization yields polyvinyl chloride or PVC or simply polythene. Ethylene can also be polymerized to form polythene.

(c) *Synthetic rubber*: From ethylene, styrene can be produced, and styrene can be used to manufacture styrene-butadene rubber (vide chapter on petroleum).

(d) *Chemical fertilizer:* Methane is the starting substance for manufacturing nitrogenous fertilizers such as ammonium sulphate, ammonium nitrate, etc. Amongst the hydrocarbons methane is the richest source of hydrogen and is the simplest to convert to hydrogen by catalytic steam reforming according to the reaction:

$$CH_4 + H_2O \longrightarrow CO + 3H_2$$

Nitogen is added to the hydrogen stream to obtain ammonia. From ammonia, the fertilizers can be manufactured.

(e) *Other chemical products:* Alcohol (ethanol or ethyl alcohol) and detergents can be produced based on ethylene, while acetone, some solvents and some drugs can be produced based on propylene. Methanol or methyl alcohol is a product based on methane.

2. Gasoline Recovery

The 'wet' natural gas contains appreciable quantities of easily liquefiable hydrocarbons —butane (C_4H_{10}), pentane (C_5H_{10}), hexane (C_6H_{14}), heptane (C_7H_{16}) and octane (C_8H_{18}). These hydrocarbons constitute the gasoline vapour and can be recovered in the liquid form. Some rich American natural gas has been reported to have yielded as much as 106 liters of gasoline per 100 m^3 of gas. The extraction nay be : (*a*) by compression with cooling, (*b*) by refrigeration or (*c*) by washing with a heavy oil and subsequent distillation.

3. Carbon Black

Carbon black is a loose amorphous form of carbon produced commercially by thermal or oxidative decomposition of hydrocarbon. It is used mainly in rubber goods (by mixing carbon black with latex), pigments and printers' ink. Natural gas is the principal feedstock from which carbon black can be manufactured by three processes as follows:

(*i*) *Contact (channel) process*: Natural gas is burnt with insufficient air. The smoke is made to strike on a cool iron channel, whereupon carbon black is deposited and is scraped out. A yield of as high as 21 kgs/1000 m^3 of natural gas has been reported.

(*ii*) *Furnace process*: A mixture of natural gas and air is fed into a reactor. Combustion of a part of the hydrocarbon raises temperature up to 1700° C, causing decomposition of the unburnt hydrocarbon to carbon black. A water spray quickly cools the hot reaction products, and the finely divided carbon black is recovered by cyclones and bag filters. A yield of as high as 260 kg/1000 m^3 of natural gas has been reported.

(*iii*) *Thermal process*: In this process, natural gas is decomposed to carbon and hydrogen by heated refractories.

4. Domestic and Industrial Heating

For this purpose natural gas can be used either directly or in the form of one of the two processed products namely liquefied petroleum product (LPP) and liquefied natural gas (LNG).

(*a*) *Natural gas*: In gaseous form, it can be supplied through pipelines to the domestic and industrial consumers. In this form, it has been used for manufacturing cement. The main disadvantages are:

- (*i*) it may prove to be cost prohibitive to construct a network of pipelines over long distances to cater to a large number of consumers;
- (*ii*) it cannot be stored easily and has to be used continuously keeping in pace with production;
- (*iii*) its supply cannot be varied in keeping with variations of demand; and
- (*iv*) it is not amenable to preheating, because its hydrocarbon constituents decompose at elevated temperature with the formation of deposits of carbon in passageways of the preheater, eventually choking the burners. However, in areas surrounding gas wells, it can serve as a good fuel for the purpose of direct firing.

(*b*) *LPG*: By compressing, the small quantities of propane and butane present in natural gas can be separated from methane and ethane in liquid form, stored in cylinders and supplied to consumers as per demand. As soon as the pressure is released, the LPG instantly gasifies so that it becomes possible to fire it for domestic cooking, metal cutting, gas carburizing, annealing, etc. In gaseous form LPG has a thermal value of 2500-3500 BthU/cft.

(*c*) *LNG*: It is also called NGL (i.e., natural gas liquid). To obtain LNG, the entire natural gas including methane and ethane is liquefied instead of only the less volatile fractions as in the case of LPG. Since methane is highly volatile and cannot be liquefied at normal temperature simply by increasing pressure, natural gas must be cooled to cryogenic temperature for liquefaction. Seventeen kilo-liters (kl) of gas at normal temperature condenses to 0.028 kl of liquid at (–) 161 °C. The advantages are:

(*i*) surplus gas can be stored during summers when demand for domestic heating is low, and then it can be easily revaporized and supplied in winters when demand is high, and

(*ii*) during revaporizing, the cryogenic temperature can be made use of in liquefaction of air for various industrial applications.

The main disadvantages are: it is extremely difficult, costly and hazardous to store a highly inflammable cryogenic liquid like LNG; a small leak in the storage tank may spread LNG which will catch fire easily, and consequently an elaborate insulation system for the tanks is a must.

5. Transportation

Use of natural gas for running vehicles like trucks, buses etc. has come out of the realm of R & D into practical reality. Since CNG mainly contains methane (CH_4) in which carbon-content is negligible compared to hydrogen, chances of CO emission is practically zero. The hydrogen serves as the fuel which yields only water on burning. Thus CNG is environment-friendly.

6. Reductant

For manufacturing sponge iron by the direct reduction technology, natural gas can serve as an effective reducing agent. In conventional blast furnace also, injection of natural gas through the blast furnace tuyeres (as a partial substitute of coke) is an established practice in some countries. In these applications, the hydrocarbons of the natural gas first undergo reformation at high temperature and then are dissociated into hydrogen and CO, both of which serve as effective reducing agents. The current of natural gas is also capable of easily pervading into the solid iron ore charge in the furnace and reacting efficiently.

7. Power Generation

Natural gas can simply be used as a fuel to generate steam for running turbines. But in a more efficient system, combined cycle technique is used. In this technique, the hot gas is first channeled through pipes into a turbine (gas turbine). The excess heat of the spent gas is then recovered for generating steam. The superheated steam is then forced to strike another turbine (steam turbine). Thus with the help of natural gas two sets of turbines can be run simultaneously and more electricity can be generated.

8. Synthetic Petroleum

Today, gas-to-liquid (GTL) is the generic name for the process of converting natural gas into a synthetic hydrocarbon liquid. The GTL products provide premium quality fuels that contain minimum toxic emissions and greater fuel economy when combusted. Through the application of technical innovations, particularly in the area of synthesis catalyst and reactor design, the processes for producing fuels from coal and natural gas are becoming more cost-effective and competitive with fuels derived from crude oil refineries.

New Zealand is the pioneering country for development of bio-technology to convert natural gas into petrol. In India, National Environmental Engineering Research Institute (NEERI) is reported to have conducted experiments for this purpose. The technology involves two operations: first, conversion of methane (CH_4) to methanol (CH_3OH), and second, transformation of methanol into petrol. Methane of natural gas can be converted to methanol by certain methane-oxidizing bacteria; but immediately after conversion to methanol, further oxidation has to be arrested, and for this purpose various reagents like cupric salts, boric acid, sodium chloride, etc., have been tried. Possibility of culturing of some suitable bacteria that can not only halt the transformation of methanol to other oxidizable products, but actually convert it into petrol, is also being investigated.

In the Fisher-Tropsch (FT) method, the production of GTL occurs in two stages. Syngas (a mixture of CO and H_2) is first produced from the natural gas within a reformer. This is then converted to synthetic crude oil 'syncrude' by a suitable catalyst such as cobalt, iron, nickel, etc., at an elevated temperature. The composition of the syncrude is controlled by the choice of catalyst and the temperature of the process. For example, 330 °C produces mostly gasoline and olefins whereas 180-250 °C produces diesel fuel and waxes. The syncrude is finally refined like its natural counterpart.

In the beginning of the 21st century, a research team in Texas, USA, has developed a radically new process for converting natural gas into hydrocarbon liquids. The process involves first converting some of the natural gas into more reactive molecules by rapidly heating the gas to a very high temperature in an electric furnace, and then passing the mixture of reactive molecules and natural gas to a catalytic reactor in which the two combine to form light natural gasoline and hydrogen. Finally, the liquid product is isolated from unreacted gas and the hydrogen. In this process the by-product hydrogen provides the required energy.

9. Helium Extraction

Helium is a colourless, odourless, inert and lighter-than-air gas. Natural gas from some fields—particularly in USA and Canada, contains economically recoverable quantities of helium.

Liquid Helium is a cryogenic substance. Its boiling point is (–)269°C and it remains liquid at the absolute temperature, i.e., 0 K or (–)273°C (cf. hydrogen solidifies at 14K). Its superfluidity below 2.2K (when most gases solidify) and also inertness make it indispensable in applications like liquefaction of oxygen for storage and transportation in cylinders, cooling of nuclear reactors, control of flow of cryogenic fuel in rockets and space shuttles, research in low temperature superconductivity, testing of refrigerators, welding, carrying of oxygen to lungs of patients, etc. In U.S.A., considerable importance is given by the Federal Government to production, stock and distribution of helium—inasmuch as a separate act called Helium Act Amendments of 1960 is in force. In accordance with this Act, the research programmes on this wonder gas are conducted and its stocks maintained to support needs for vital research activities in the fields of space and defence and also in the universities.

10. Re-pressurizing Oil Reservoir

This is one of the earliest uses of natural gas. It has long been realized that the pressure under which oil and gas exist in an underground reservoir, together with the amount of gas actually dissolved in the oil itself, provides the propulsive power to force the oil up through the wells. It is desirable, therefore, that the gas/oil ratio in the reservoir remains high. This ratio, however, decreases as gas keeps escaping through the walls along with the recovered oil. To restore the gas/oil ratio in the reservoir, it may be necessary to pump the escaping gas back to the reservoir.

11. Fuel Cell

It is believed that fuel cells will become the norm in many applications within the first quarter of the 21st century—be it for transport or energy supply or in an industrial application. Fuel cells can use a variety of fuels like natural gas, petroleum, methanol, hydrogen, etc., to produce electricity through a noncombustion electrochemical reaction. The fuel used directly is hydrogen, usually reformed from hydrocarbon fuels. A catalyst splits the hydrogen molecules into electrons and protons. The protons pass through an electrolyte membrane, while the electrons create an electric current. The electrons and protons are then re-united and combined with oxygen to create water. The process also creates heat. Several types of fuel cell technology are in various stages of development. They are:

(*i*) Phosphoric acid fuel cell
(*ii*) Molten carbonate fuel cell
(*iii*) Proton exchange membrane fuel cell
(*iv*) Solid oxide fuel cell
(*v*) Alkaline fuel cell
(*vi*) Direct methanol fuel cell
(*vii*) Regenerative fuel cell
(*viii*) Hydrogen peroxide fuel cell.

Out of these, the Phosphoric acid and the Alkaline cells are already in use. The former type of cells are installed at utility power plants, hospitals, hotels, schools, office buildings and airport terminals; and the latter type is long used by the National Aeronautics and Space Administration (NASA) in space missions.

SPECIFICATIONS OF USE

There is no specification as such except that sulphur is considered to be a deleterious constituent in most of the uses of natural gas.

For manufacturing of various chemical products, for recovery of gasoline, for use as a reductant, and for conversion to synthetic petroleum, chemical composition of natural gas particularly the hydrocarbon content is important, while for manufacturing carbon black, the carbon content assumes importance. In domestic and industrial heating, both thermal value and chemical composition (hydrogen content) assume significance, and for use in the form of LNG, the gaseous form (high volatility) is also important. For use in transportation also (in the form of CNG), the hydrogen content or in other words, methane content, should be high. In power generation, the kinetic energy of hot gas is used to move one set of turbines while the thermal value is needed for steam generation to move another set of turbines; for re-pressurizing oil reservoirs, the propulsive power of the gas is harnessed. Therefore, in application for power generation, both the thermal value and the gaseous form of natural gas should be the deciding factor, while in that for re-pressurizing oil reservoirs the gaseous form alone is enough. For recovery of helium, the helium content should be high.

UTILIZATION OF WASTES

Earlier, natural gas itself used to be regarded as a waste byproduct in petroleum recovery, because, on one hand it could not be utilized in industries due to lack of costly storage and transportation facilities, and on the other hand its escape reduced the gas/oil ratio in the oil reservoir and consequently the recovery of oil itself. Considerable quantities of natural gas escaping through oil wells used to be simply flared. During the recent years, however, natural gas has come to be regarded as an economic commodity rather than a waste product. Thus all the modern industrial uses of natural gas themselves exemplify utilization of a so called waste product.

However, though the chemical components of natural gas are extracted for economic use, sulphur is regarded as a deleterious constituents in most of the applications. The common form in which it occurs is H_2S gas. Now-a-days, the sulphur is recovered by oxidizing the H_2S.

SUBSTITUTION

Natural gas itself is tending to substitute some traditional commodities like coal and petroleum, in uses as a fuel for heating purpose, for transportation, for power generation and in extraction of chemicals. However, in certain uses, natural gas can be substituted by some other substances.

1. Reductant

In manufacturing of sponge iron through direct reduction process, non-coking coal can be used as a reducing agent in areas where natural gas is either costly or not available.

2. Carbon Black

In the furnace process of manufacturing of carbon black, petroleum can be used as the feedstock instead of natural gas. In USA, the trend is to use more of petroleun for this purpose.

3. Methanol

In production of methanol—for eventual conversion either to synthetic petroleum or to other chemical products, 'synthesis gas' or 'syn-gas' can substitute natural gas as the starting raw material. Syn-gas is a mixture of CO and H gases, and in presence of some catalyst it can be reduced to methanol (CH_3OH) as follows:

$$CO + 2H_2 \longrightarrow CH_3OH$$

4. Methane hydrates

Methane hydrates containing 90% methane have been found to occur dissolved in sea water under certain conditions. These hydrates are formed from gas and water at low temperature and high pressure prevailing at depths ranging from 300-500 meters below sea surface at some places. These might become an important source of methane in future, and are receiving attention of the scientists since the beginning of the 21st century.

5. Solar Hydrogen

Australian scientists predict that a new way to harness power of the sun to extract clean and almost unlimited energy supplies from water will be a reality in the near future. According to them, using special titanium oxide ceramics that harvest sunlight and split water, it will be possible to produce hydrogen fuel in an environment-friendly manner.

5

CHAPTER

URANIUM ORE AND METAL

Uranium is the heaviest naturally occurring element. Its atomic number is 92, i.e., there are 92 protons in an atom of uranium, while the number of neutrons may be either 143 or 146. Thus the atomic weight of uranium may commonly either be 235 (i.e., 92 + 143) or 238 (i.e., 92 + 146), signifying two common isotopes of uranium—U_{235} and U_{238} (isotopes of an element have the same number of protons, but different number of neutrons). A third isotope—U_{234} is also found in nature, but rarely.

The earth's crust contains on an average about 4 ppm of uranium (cf. concentration of silver is about 0.3 ppm). But the most common ore of uranium is pitchblende. It is the massive, amorphous or cryptocrystalline variety of uraninite. It is believed that the ideal composition of uraninite and pitchblende is UO_2. But the mineral is always found in nature in oxidized form, and in association with various gangue matter, so much so that no definite formula can be attributed to it. But the naturally occurring pitchblende mainly contains U_3O_8.

In Utah and Colorado areas of USA, another variety of ore of uranium, namely carnotite is found. Carnotite is a secondary mineral, produced by the natural leaching action of acid and alkaline groundwaters, and it occurs intermixed with sandstones. Its formula is $K_2O.2UO_3.V_2O_5.3H_2O$. It may contain up to 7% uranium.

In addition to conventional uranium deposits, vast quantities of unconventional resources exist. Such unconventional resources include phosphate deposits with average uranium concentrations between 100 ppm and 200 ppm in sedimentary rocks, and in exceptional conditions more than 1000 ppm in igneous rocks. In USA, extraction of uranium from phosphoric acid by liquid-liquid separation process using solvent, has been demonstrated on an industrial basis. Other unconventional uranium resources explored and attempted for extraction in the past are black shale deposits and granite rocks with elevated uranium concentrations—but only on experimental scale. Further, the vast amount of uranium contained in seawater (concentration about 3 ppb) is often seen as eventual "back-stop" uranium resource; but prohibitively high recovery costs compared to conventional alternatives have stopped all investigations that were taken up in the 1970s. There are many other rarer uranium ores, which are also not of much wide scale economic relevance. But owing to bioleaching technology, mined out stopes in many old underground uranium ore mines and also very low grade ores which were previously considered as uneconomic are now becoming economic resources of uranium.

In India, pitchblende is the main uranium ore being mined, though monazite sands also contain some U_3O_8. The discussion on ore, therefore, will be confined to pitchblende mainly.

HISTORY

The history of analytical and mineralogical study of pitchblende dates back to early eighteenth century, in Germany. Though the name 'pitchblende' gained currency in Germany by 1758, still the scientists were not sure about its chemical composition, and it used to be regarded as an ore of zinc.

It was in 1789 that a German chemist, Martin Heinrich Klaporth discovered the presence of a new substance in the pitchblende from Joachimsthal deposit in Germany (now in Czecholovakia). He named it 'uranit', to commemorate the planet Uranus which had been discovered about 8 years earlier. Uranit was a compound of a metal which Klaporth termed as uranium. However, his attempts (as well as subsequent attempts by other scientists for the next 50 years or so) to produce uranium metal by reducing uranit was unsuccessful. All he could produce was UO_2. The metal uranium was separated for the first time in 1841, by a French chemist, Eugene Melchior Peligot.

During more than 100 years since its discovery in 1789, uranium was recorded as a worthless metal. In 1896, Antione Henri Becquerel discovered that the radiation emanating from uranium salts would darken film. Marie Curie, in the same year, established that the radiations came from uranium itself, and she gave the name 'radioactivity' to this phenomenon. This discovery opened the door to radioactivity studies. In 1898, Pierre Curie discovered another radioactive element, radium (atomic weight 226). In fact, radium is an intermediate radioactivity decay product of uranium, and uranium ores expectedly contain small amounts of radium. Since radium is more radioactive than U_{238}, the experimental rsearches in radioactivity were mainly confined to this element during the early twentieth century, and uranium ores were of interest for the radium they contained. Apart from this academic interest, the only commercial interest that the uranium salts generated during those years, was minor application in metallurgy (as an additive to certain alloys), glass (coloured fluoroscent glass), textile industries and photography.

These researches in the field of radioactivity, however, were marked by a few revolutionary milestones. First of all, in 1905, Albert Einstein theoretically propounded the famous 'special theory of relativity'. According to this theory, instantaneous radioactive decay would yield enormous energy. He mathematically derived the famous formula:

$$E = mc^2$$

Where 'E' is the energy released,

'm' is the mass of the atom

and 'c' is the speed of light (300,000 km/sec)

The second milestone was achieved in 1919, when Lord Rutherford was for the first time, able to split the atom of nitrogen by bombarding it with alpha particles. The third landmark was James Chadwick's discovery of neutron in 1932. This discovery was actually the logical end of the experiments of Irene Curie and Federick Curie Jolio during 1931-32. They hit boron atom (B_{10}) with helium (He_4 or alpha particles) resulting in release of what Chadwick confirmed as free neutrons.

Finally, the real turning point came in 1938. In that year, two German Scientists, Otto Hahn and Fred Strassman were for the first time able to split U_{235} atom by bombarding it with neutrons. This was followed by the achievement of chain reaction in 1942 by an Italian scientist Enrico Fermi. The first atomic fission bomb using uranium was exploded in 1945, i.e., towards the end of World War-II. This period marked the birth of atomic age, opening up the tremendous possibilities of both constructive and destructive use of uranium. Only then onwards, uranium ore started being mined for the uranium it contained, and not just for the radium.

The first uranium mining took place in the middle of nineteenth century in Joachimsthal (then in Germany, now in Czecholovakia). Subsequently, Cornwall district in UK, Colorado state in USA (1896), Ferghana in Russia (1908), Sweden (1909), Urgeirica in Portugal (1911), Radium Hill in Australia (around 1915), Shinkolobwe in Zaire (1921) and Eldorado in Canada (1930), entered into the uranium map of tho world. It has been estimated that the total uranium production prior to the discovery of fission in 1938 was only about 7,500 tonnes. But the real impetus to uranium mining was received after World War-II, when extensive uranium exploration programmes were initiated throughout the world, resulting in a spate of discoveries of uranium deposits.

In India, occurrences of uranium mineral was recorded for the first time in 1860 by Emil Stoehr. The mineral was torbenite, i.e., hydrated phosphate of uranium, which was found in Bihar. The mineral pitchblende was first reported in 1901 by Thomas Holland, who found it associated with the mica pegmatites in Bihar. However, researches in the use of uranium in India for atomic energy had to wait till 1948, when the Atomic Energy Commission was created. The architect of development of nuclear technology in India was Dr Homi Jahangir Bhabha. Under his guidance, the first reactor 'Apsara' was commissioned in 1956 for generation of power, and this heralded the atomic age in India.

CRITERIA OF USE

1. Chemical Composition

The usable fuel in natural uraninite (U_3O_8) is either metallic uranium (U_{238} and U_{235}) or UO_2. Very rich ores, such as the substantially pure uraninites of Zaire and Spain, that were mined in the beginning, contained as high as 60-70% U. However, generally the U_3O_8 content in the natural ores is of the order of 0.5-1.0%, whereas the U-content may be of the order of 0.1-0.2 percent. The natural uranium comprises 0.0056% of U_{234}, 0.718% of U_{235} and 99.276% of U_{238}. Of these, U_{235} is very important commercially. U_{238} is also of much economic significance.

Because in most ores, the uranium is combined with many elements, the ore-processing required to obtain UO_2 or U-metal is very extensive. The low grade of natural ore necessitates upgradation by concentration. The steps are as follow:

(*a*) The ore is first crushed and subjected to physical processes such as flotation and heavy media separation.

(*b*) It is then leached in either dilute H_2S_4 (acid leaching) or in an aqueous solution of Na_2CO_3 (alkaline leaching). Precipitation from the resultant concentrated liquours is achieved by adding NH_3 or MgO or NaOH. The products are either ammonia diuranate or magnesium diuranate or sodium diuranate, which constitute the input for the extraction of metallic uranium and its pure compounds.

The concentrates of these diuranates may contain 50-90% U_3O_8.

The next stage is refining of the concentrate. The steps are as follow:

(*a*) The concentrate is first redissolved in nitric acid.

(*b*) An aqueous solution of high purity uranium compounds is obtained by solvent extraction process. The solvent used is tributyl phosphate dissolved in kerosene or hexane. The product is pure uranyl nitrate solution.

(*c*) Uranyl nitrate is thermally decomposed to obtain UO_3, which is then reduced to UO_2 by hydrogen.

(*d*) UO_2 is heated in presence of anhydrous HF gas to yield UF_4.

(*e*) UF_4 is reduced to metallic uranium by reacting it with either calcium metal or (more widely) magnesium metal. The reaction is:

$$UF_4 + 2Mg \longleftrightarrow U + 2\,MgF_2$$

(*f*) Sometimes it may be necessary to enrich the concentration of U_{235} in the metallic uranium or its compounds. For this purpose UF_4 is first fluorinated further to yield UF_6. The UF_6 is subjected to differential gaseous diffusion through microporous filters. The lighter U_{235}-containing UF_6 (molecular weight 349) and the heavier U_{238}-containing UF_6 (molecular weight 352) are partially separated. In this way the concentration of U_{235} may increase from about 0.7% to 3 per cent. The enriched UF_6 so obtained is reduced again to UF_4 by means of hydrogen. Enriched uranium metal can then be obtained by metallic reduction of UF_4 with Ca or Mg metals.

Enriched uranium, enriched UO_2 and also natural uranium are the key in the principal uses of uranium ore.

2. Radioactivity

Radioactivity is the spontaneous disintegration of certain heavy elements accompanied by the emission of high energy radiation, which consists of three kinds of rays: 'alpha particles', 'beta particles' and 'gamma rays'. Alpha particles are made up of two protons and two neutrons and are positively charged. In effect, they are nothing but the nuclei of helium atoms. Beta particles are made up of electrons and are negatively charged. Gamma rays are electromagnetic rays like X-rays, travelling at the speed of light, and they carry no charge. These rays have much more penetrating power than alpha and beta particles, and can penetrate most of the metals. The ultimate end product of radioactive disintegration is one of the isotopes of lead. All radioactive phenophena die away after a certain length of time. The period in which the number of atoms of a radioactive substance decreases to one half its original value (with proportional increase in the mass of lead produced) is called 'half-life'. The rates of natural decay of the radioactive elements vary widely. For example, the half life of U_{238} is 4,500 million years, that of radium is 1,600 years and that of plutonium is 24,000 years, while that of the unstable U_{239} is merely 23 minutes.

It is the gamma ray emitted due to the radioactivity of uranium that makes uranium useful in many applications.

3. Fission

In the structure of an atom, the role of neutrons is very important. When there are two or more positively charged protons present, they tend to repel each other due to similar charge, and the nucleus of the atom tends to split. However, the neutrons hold the protons together. If the number of protons in an atom happens to be too large and the number of neutrons happens to be insufficient (i.e., in other words, if the atomic number happens to be too high and the atomic weight is not high enough), then the atom may split on its own. In reality, however, varying degrees of external energy is required to be applied in order to split the nucleus of an atom. The higher the atomic number, lesser will be that external energy required, and for the same atomic number, lower is the atomic weight, the easier it will be to split the atom. The external energy for this purpose is provided partly by the kinetic energy of a free neutron which is bombarded on the atom and partly by the absorption of that additional neutron by the bombarded atom. This ability of an atom to split due to collision with a free neutron is called 'fission'. Uranium having the highest atomic number needs relatively low level of kinetic energy for producing nuclear fission, while the kinetic energy required to produce fission in elements lighter than tantalum is too great to carry any practical sense.

Energy-wise, neutrons are broadly classified into two categories—fast and slow (or thermal). Fast neutrons have energies in the range of 1 MeV to 100 KeV (million electron volts and kilo-electron volts respectively), while the slow or thermal neutrons have energies of about 0.025 eV (one electron volt is equivalent to 1.6×10^{-12} erg). Slow neutrons are called *thermal neutrons* because the energy of 0.025 eV is the energy of a neutron in thermal equilibrium with its environment.

U_{235} is more susceptible to fission than U_{238}, because of the latter's higher atomic weight. Fission in U_{238} atom requires 5.9 MeV of energy, out of which 5.3 MeV is generated due to assimilation of an additional neutron, while the balance 0.6 MeV has to be derived from the kinetic energy of that neutron. Thus a fast neutron is needed for fission of U_{238}. On the other hand an atom of U_{235} needs only 4.5 MeV of energy for fission. Since more than this is generated by mere assimilation of an additional neutron, its kinetic energy is immaterial, and thus, thermal neutons can serve the purpose. As a result of fission, a fissile nucleus breaks into a pair of unstable fragments, which undergo further radioactive decay until stable fission products are formed. But in the process some free neutrons are also released from the original nucleus, and these free neutrons become available for producing fission in more atoms. Under favourable conditions, this process may continue endlessly to produce what is known as chain fission or chain reaction. There may be a wide range of fission products of U_{235} such as barium and krypton, cesium and ruthenium, etc.

Another result of fission is that the original fissile nucleus loses a portion of its mass, which is transformed into energy according to Einstein's equation: $E = mc^2$.

4. Phase Transformation and Alloying Behaviour

Uranium metal exists in three distinct allotropic modifications, depending on temperature. The first phase namely alpha phase changes to beta phase at an average of 661.1 °C; the beta phase changes to gamma phase at an average of 768.8 °C; and finally the metal melts at an average temperature of 1129.7 °C. These temperatures, however, are for uranium containing 50-100 ppm of impurities and may vary according to the rates of heating and cooling. The transformation kinetics of uranium has an important bearing on its alloying behaviour. For an effective and stable alloy system it is, inter alia, necessary:

(*i*) that the rate of atom migration in solid solution should be high,

(*ii*) that there should not be marked tendency for intermediate compound formation,
(*iii*) that in one or the other phase the solubility should be high, and
(*iv*) that on cooling, fine sized grains should be produced.

There is a number of metals with which uranium is capable of forming effective alloys.

5. General Physical Properties

Freshly prepared untarnished surfaces of uranium metal have the silver white colour, metallic luster and high thermal conductivity. It is opaque, malleable and a fair electrical conductor. Pitchblende is usually black, dark grey or brown in colour. Some uranium compounds are green in colour and they fluoresce under ultraviolet light.

COMMON USES OF ORE AND METAL

The common uses of uranium ore, metal and compounds are as follows:

1. Nuclear power generation
2. Submarines and ships
3. Atom bomb or fission bomb
4. Uranium alloys
5. Chemicals and compounds
6. Sterilisation and radiotherapy
7. Geological age determination
8. Photography
9. Glass and ceramics
10. Textile and leather
11. Incandescent light
12. Malaria control

These uses are now discussed as follows:

1. Nuclear Power Generation

In this application, uranium is used to produce heat energy which in its turn can be used to generate steam in boilers and then to move turbines as in the case of thermal power generation with the help of coal (see chapter on coal). For production of heat energy, the property of fission of uranium is the principal criterion, while its high thermal conductivity is also beneficial.

(*a*) *Fuel*: The first step is to select the nuclear fuel. Since U_{235} is more fissile than U_{238}, the fuel should contain sufficient concentration of U_{235}. Natural uranium contains about 150 parts of U_{238} for every part of U_{235}. This is not adequate to sustain a chain fission. When U_{235} atom is bombarded with a neutron, it produces energy and also releases two neutrons, which in turn hit two more atoms of U_{235}, produce more heat and release 4 neutrons which then can hit 4 more U_{235} atoms, and the fission continues. If however there is insufficient number of U_{235} atoms surrounded all around by U_{238} atoms, then many of the released neutrons will hit U_{238} atoms instead of U_{235} atoms. Since U_{238} atoms are ordinarily not very fissile, a stage will soon come when the fission comes to a stop and so does generation of heat. To pre-empt this possibility the U_{235} in the fuel is raquired to be enriched to about 3 per cent, and this enriched fuel has been found suitable for chain reaction to sustain in a controlled manner long enough to produce sufficient heat. The material used may be:

— massive uranium metal

— uranium alloy such as with aluminium, boron, beryllium, bismuth, copper, lead, manganese, molybdenum, nickel, niobium, silicon, tin, titanium, zirconium

— uranium compounds such as UO_2, UC, UN.

Uranium metal and UO_2 are, however, more commonly used than the other types of fuel.

(b) *Fuel element*: Fuel element is a sort of container or matrix within which the fuel is placed. In addition to the normal physico-mechanical properties of uranium, the changes taking place during fission have also to be taken into acount in the design of a fuel element. When most metals are used in a high temperature environment, the heat transfer is from outside into the metal, and the temperature gradient will be small within the metal once equilibrium is reached. In case of uranium fuel, however, the heat is generated by fission within the fuel, and it flows from within outwards. Besides, radiation from the fission product and oxidation of the fuel by coolant materials (like CO_2) have also to be taken into account while selecting the material for fuel element. Magnesium-beryllium alloy, pure magnesium, pure aluminium, zirconium, glass, etc., are some of the materials used for fabricating fuel elements. The fuel element may be of two types—container type and dispersion type. In the container type, the fuel embodied within the container is generally in the form of plates, sheets, bars, billets, rods, tubes and bunches of wires. It can also be in the form of pellets of UO_2 packed inside stainless steel or beryllium tubes. In the dispersion type of fuel elements, the fuel (uranium metal or compound) is distributed in discrete particles throughout a matrix of metal, glass etc., with which the particles must not chemically react at the operating temperatures.

The container or matrix in a fuel element should, inter alia, have low neutron absorption (so that neutron economy of the fuel is maximized) and good thermal conductivity (so that the heat energy generated due to fission within the fuel is transferred efficiently).

(c) *Reactor*: The fuel elements are housed in a 'reactor'. The function of a reactor is primarily (*i*) to contain the dangerous radioactive emissions, and (*ii*) to prevent loss of the heat generated within it. Reactors may be of two types: (*a*) thermal reactors which use thermal or slow neutrons (0.025 eV energy), and (*b*) fast or fast breeder reactor, which use fast or high energy neutrons (1 MeV-100 KeV) and in which more fissionable material is produced than consumed. The reactor walls consists of thick schields of steel and concrete. Within a reactor, mechanisms are provided:

(*i*) to extract the generated heat for producing steam to move turbines,

(*ii*) to regulate the generation of heat depending upon requirements, and

(*iii*) to control the speed of the flying neutrons freed due to fission.

For extracting the heat, coolants like gas (CO_2), ordinary water (H_2O) and heavy water (D_2O) are used. For regulating the heat, control rods made up of materials like boron, cadmium, etc., are used. These materials have the ability to absorb neutrons. If the rods are lowered down in the reactor, more neutrons will be absorbed and the process of fission will slow down, thus reducing the heat. On the other hand, if the rods are pulled up out of the reactor, then less number of neutrons will be absorbed, and consequently process of fission will accelerate, thus increasing the heat. For controlling the speed of the neutrons, moderators are used in thermal reactors, that use low energy neutrons for chain reaction. The moderator should only bring down the energy of the incident neutrons, and should not absorb them. Heavy water and graphite make good material for moderator.

It can be calculated that according to Einstein's equation $E = mc^2$, splitting of one atom of U_{235} generates 200 MeV energy (as a result of mass loss), and that one gramme of U_{235} can produce 23000 units of electrical energy as shown below:

200 MeV = $200 \times 1.6 \times 10^{-6}$ erg = 3.2×10^{-4} erg

1 gramme-atom of U_{235} contains 6.02×10^{23} atoms

1 gramme-atom of U_{235} = 235 grammes

& 1 erg = 10^{-7} watt-seconds

So, 235 grammes of U_{235} generate $3.2 \times 10^{-4} \times 6.02 \times 10^{23}$ ergs

Or 1.93×10^{20} ergs or 1.93×10^{13} watt-seconds

So, 1 gramme of U_{235} generates $(1.93 \times 10^{13})/235$ or 8.2×10^{10} watt-seconds or 23000 kilowatt-hours energy.

Generation of the same amount of electricity will require about 3 tonnes of coal.

2. Submarines and Ships

In maritime transportation, nuclear powered engines are used. Essentially miniature reactors are installed and the electricity thus generated is used as the motive energy. The advantage of uranium as fuel is that it occupies very small space, unlike diesel which occupies large storage tanks to sustain long voyages. In case of submarines, there is an additional advantage. Diesel engines need air and batteries need recharging. So submarines powered by diesel or battery cannot remain under water for long. On the other hand, nuclear powered submarines do not suffer from this problem.

3. Atom Bomb or Fission Bomb

In this use also, the amenability to fission is the key criterion and so uranium containing enriched U_{235} is used, as in the case of power generation. However, in reactors, the chain fission takes place in slow and controlled manner, whereas in atom bombs, chain fission should proceed very fast and high level of heat energy is required to be generated instantaneously. This not only requires a high degree of enrichment of the U_{235}, but also adds importance to the size of the uranium mass.

If a spherical mass of U_{235} is small, more neutrons are lost from the surface than are retained in the volume, and hence self-supporting chain fission cannot be maintained. If the mass is increased, more neutrons will be retained inside and the chain fission will continue longer. Eventually, 'critical' mass is reached which just supports an uncontrolled chain fission.

This size is called 'critical size' and the size just smaller than this is called 'subcritical size'. Though the critical size is a top secret formula, it is believed to be between 1 and 100 kgs.

Apart from the problem of high degree of enrichment and the formula for critical size there is a third problem in the actual making of an atom bomb. There should be a mechanism by which the bomb should not explode after its manufacture till it is transported to the appointed place at the appointed time. Theoretically, this is possible if the bomb consists of two subcritical sized uranium masses separated by a partition, and that partition can vanish by remote control or by melting or otherwise, sometime after it is dropped. As soon as the partition vanishes, the two subcritical sized components merge together and become supercritical sized, resulting in uncontrolled chain reaction and instantaneous release of heat.

4. Uranium Alloys

(*a*) *Uranium alloys suitable as fuel*: Some of the uranium alloys have been found to be good fuel, because of advantageous properties. These are:

- Uranium-aluminium (aluminium has low neutron absorption and it imparts corrosion resistance)
- Uranium-antimony (antimony has low melting point and this alloy is suitable in liquid fuel reactors)
- Uranium-bismuth (same as antimony alloy)
- Uranium-boron (B_{10} isotope has a high neutron absorption and so is not suitable as a component in fuel; however B_{11} isotope has low neutron absorption and is suitable as a dispersant in the dispersion type fuel element)
- Uranium-magnesium (magnesium has high thermal conductivity and is resistant to radiation damage; this alloy is suitable as a matrix in dispersion type fuel element)
- Uranium-molybdenum (suitable as a fuel in both thermal and fast reactors)

(*b*) *Other uranium alloys*: The other alloys include: uranium-beryllium, uranium-chromium, uranium-cobalt, uranium-copper, uranium-gallium, uranium-germanium, uranium-gold, uranium-indium, uranium-iridium, uranium-iron, uranium-lanthanum, uranium-lead, uranium-manganese, uranium-mercury, uranium-molybdenum-niobium, uranium-molybdenum-plutonium, uranium-molybdenum-ruthenium, uranium-molybdenum-titanium, uranium-molybdenum-zirconium, uranium-nickel, uranium-niobium, uranium-niobium-zirconium, uranium-neptunium, uranium-palladium, uranium-platinum, uranium-ruthenium, uranium-silver, uranium-tantalum, uranium-tellurium, uranium-thorium, uranium-tin, uranium-titanium, uranium-tungsten, uranium-vanadium, uranium-zinc, uranium-zirconium.

However, the commercial potentiality of these alloys is not well understood.

5. Uranium Chemicals and Compounds

Oxide (UO_2), carbide (UC & UC_2), nitride (UN) and sulphide (US & US_2) are suitable as fuel in dispersion type fuel element. UO_2 is also suitable in container type fuel element.

U_3Si has also been tried as a fuel. Though it is fairly resistant to both corrosion and radiation, it is not amenable to casting. Besides, compounds of hydrogen and phosphorus have also been made, though they do not appear to be of any practical significance.

Amongst the uranium salts, $Na_2U_2O_7$, UO_2Cl_2, $(NH_4)_2U_2O_7$, uranyl sulphate and uranyl nitrate, sodium uranyl carbonate, uranyl acetate and various uranates are common.

6. Sterilization and Radiotherapy

The property of radioactivity—more particularly the ability to emit gamma rays, is the key to the use of uranium in this area. Gamma rays are harmful to life, but if suitably regulated, they can kill germs as well as create conditions in which germs cannot thrive for a considerable length of time even after withdrawal of the rays.

Uraniun, however, has a very slow rate of emission of gamma rays, and so it is not suitable for this use directly. To overcome this problem natural cobalt is inserted into reactors. It absorbs neutrons and the isotope cobalt-60 is formed. This cobalt-60 has the ability to emit gamma rays much faster, and can be conveniently used for the purpose of sterilization and radiotherapy.

Items of food, seeds, surgical instruments may be sterilized by exposing them to gamma rays emitted by cobalt-60. In the same way, harmful microbes in diseased body cells can be controlled through radiotherapy.

7. Geological Age Determination

In this application also, the property of radioactivity of the uranium is the key. Uranium-238 decays at a fixed rate to its ultimate stable end product lead-206. Since, this rate (i.e., half-life) is known, then by measuring the ratio of the atoms of U_{238} and those of lead-206 in a rock, it is possible to determine the age of that rock.

8. Photography

Photographic plates are sensitive to gamma rays (as they are to X-rays). This is taken advantage of in detecting damaged cells in human body. If a radioactive element is inserted into the region affected by the damaged cells through some suitable chemical carrier, then gamma rays will be emitted from that region. If a photographic film is exposed to these gamma rays, then the affected region will show as patches on the photograph. However, uranium is not directly used in this manner.

9. Glass and Ceramics

In these uses, uranium salt is used. In manufacturing of glass, it is used as an additive to make fluoroscent glass of an opalescent yellow transparency, which turns green in reflected light. In ceramics, it is used to impart pale greenish yellow colour to the glazing material.

10. Textile and Leather

In this case also uranium salt is used. Its function is to fix the colouring material in calico-printing and dyeing.

11. Incandescent Light

For this purpose, uranium ore was used during the earlier part of its history. Radium is an intermediate decay product of uranium:

$$U_{238} \longrightarrow \text{Radium-226} \longrightarrow \text{Radon-222} \longrightarrow \text{Lead-206}$$

So, natural uranium ore can be expected to contain some traces of radium. It is this radium in uranium ore which produced the incandescent light. Incidentally, radium is valued for luminous painting of clock dials.

12. Malaria Control

In the beginning of the 21st century, the United Nations Organization has launched a project to harness nuclear technology for eradicating the mosquitoes whose bite transmits malaria, a deadly disease devastating the African continent. In the 'Sterile Insect Technique (SIT)' as it is called, it is envisaged that mosquitoes will be bred and the males will be exposed to enough radiation to render them sterile; the males will then be released into the environment to breed with the females, whose eggs will remain unfertilized and will never hatch. The whole concept is that the mosquito population will start to crash and eventually may actually lead to eradication of the insect and with it, the disease.

UTILIZATION OF WASTE

Wastes may generate in reactors, in uranium concentrator plants as tailing, and also due to obsolescence of some particular use. These are discussed as follows:

1. Waste in Reactors

There is a strong public concern about the lack of adequate methods for disposal of nuclear wastes. As in the beginning of the 21st century, a number of countries (e.g., Denmark, Italy, Austria, Sweden and Germany) have opted not to construct new nuclear power plants and for the phase out of current plants. Two types of end products can generate in a reactor as a result of chain fission of uranium. Uranium contains both U_{235} and U_{238}, and the end products differ accordingly.

(*a*) *Uranium-235*: Due to splitting of the atoms, two end products are produced. It is not certain, which specific products are formed, and there may be a variety of pairs of elements. It is believed that the most probable products of thermal neutron fission are those with mass numbers 95 and 139. But the spent fuel remains radioactive for a long time. While presently there is no use for the spent fuel, still it has to be disposed in a manner so that no harm is caused by its radioactivity. Various methods have been tried including sealing it in a container and burying it underground or under the ocean, and also converting it into solid glass which can be stored more easily until it ceases to be radioactive. Based on laboratory investigations a bioflocculant has been isolated from the seeds of a forest tree 'stychnos potatorium'. This bioflocculant is claimed to have capability of absorbing uranium and of cleansing nuclear waste.

(*b*) *Uranium-238*: As a result of bombardment of a U_{235} atom by a neutron, two neutrons are freed. One of them is required to hit another U_{235} atom. The other one may hit one of the surrounding U_{238} atoms. The latter, on absorption of a neutron, is converted first to U_{239}, which being highly unstable (half life 23 minutes) quickly changes to neptunium-239 (Np_{239}) through emission of beta particles. This Np_{239} is also unstable (half life 2.3 days) and it keeps emitting beta particles till finally it is converted to stable plutonium-239 (Pu_{239}). Thus, a large quantity of plutonium is produced at the end of a fission process in a reactor. This Pu_{239} is amenable to fission by high energy fast neutrons. It can be separated from the rest of the end products in a reactor by chemical methods, and can be used as a fuel in a fast reactor.

2. Tailings from Concentrator

In India, there is a uranium mine in Jadugoda. The uranium ore contains almost equal quantities of uraninite (about 3%) and magnetite (3.2%). The magnetite goes into the concentrator plant tailings, in which the content of 'magnetics' is about 3 percent (magnetics together with some nonmagnetic constituents like FeO, Fe_2O_3 and SiO_2 make up the nineral magnetite). The tailings are treated to recover magnetite containing over 80% magnetics. This recovered magnetite, which is in granular form, is suitable as a heavy medium in coal washing, and is actually used along with natural magnetite grains.

3. Obsolescence of Use

This is an unusual development due to the political environment of the world. Over the years nuclear arsenals (atom bombs, missiles etc.) have been stockpiled by countries like USA, Russia as a measure of preparedness for possible war. Now, due to easing of international tension and due to a strong campagn for nuclear disarmament, these arsenals and large quantities of highly enriched uranium have become redundant. The question of effective utilization of this highly enriched uranium needed to be addressed. USA has developed a process by which the highly enriched uranium can be converted back into low enriched uranium, so that the latter may be used in commercial power reactors for generation of electricity.

SUBSTITUTION

In power generation, conventional fuels like coal, natural gas, etc., and various sources of non-conventional energy like solar energy, wind energy etc. can always wholly or partially substitute uranium and vice versa. But some superior substitutes for uranium in different uses are as follows:

1. Plutonium

It has been mentioned that Pu_{239} is an end product in thermal reactors, which can be reused in a fast reactor. But now-a-days a special type of fast reactor has been developed which only at the starting point needs this 'waste' from a thermal reactor; thereafter, it produces this fuel on its own. In other words, this plutonium which is bred due to its own fission, is a sort of substitute of U_{235}-based fuel, because the fission of the latter is no longer necessary for Pu_{239} to be produced. This type of reactor is called 'fast breeder reactor'.

In a fast breeder reactor, a mixture of Pu_{239} and U_{238} is used as a fuel (the initial charge of plutonium is obtained from the spent fuel of thermal reactor). The fission of Pu_{239} releases free neutrons, some of which are absorbed by the U_{238} to produce Pu_{239}, and thus not only the consumed plutonium is replenished, but some surplus may also be generated. In effect, the reactor continues to operate without requiring any fresh charge of plutonium (only the U_{238} needs to be replenished from time to time).

In India, a unique fuel for fast breeder reactors has been developed. It consists of a mixture of 70% plutonium carbide and 30% uranium carbide, made in the form of pellets.

Fast breeder reactors have, however, some problems—particularly in the area of controlling the fast generating heat. Liquid sodium has been tried as a coolant, because it has excellent heat transfer characteristics, and there is a wide difference between its melting point (98 °C) and boiling point (882 °C). On the other hand, it is dangerous to handle liquid sodium because

(i) the hot metal burns freely when exposed to air,

(ii) it reacts explosively on contact with water, and

(iii) it is highly corrosive and requires special alloys for making storage tanks and circulation pipes. Working with plutonium in a fast breeder reactor, therefore, requires a high degree of technological and operational expertise.

2. Thorium

Th_{232} can also be used in much the same way as Pu_{239}. This is discussed in the chapter on 'Monazite and Thorium'.

3. Fusion Bomb

This bomb is a much superior substitute of atom bomb (or fission bomb). The underlying principle is that when one atom penetrates into another atom and becomes one fused atom, the mass of the new atom is less than the combined mass of the old atoms. Due to this loss of mass, energy is released as per Einstein's equation: $E = mc^2$. This principle has been applied to make bombs, by fusing deuterium atoms. Since deuterium is an isotope of hydrogen, such bombs are called 'hydrogen' bomb. Also, since tremendous amounts of heat is generated in such bombs due to nuclear fusion, these are also called 'thermonuclear' bombs.

Though, no radioactive waste is generated due to such fusion, it has not been possible to take advantage and apply this principle for generating power, because scientists have not succeeded in effecting fusion in a regulated manner.

6

CHAPTER

LIGNITE

Lignite or brown coal is regarded as the lowest rank coal. It is an intermediate stage after peat and before the formation of bituminous coal in the coalification process.

CRITERIA OF USE

Chemically, it contains low ash (of the order of 3-14%), low fixed carbon (of the order of 30-45%), fairly high volatile matter (of the order of 25-55%), and high moisture (of the order of 10-20%, when air dried). It contains some decomposed vegetable matter. Its thermal value is lower than coal and is of the order of 9,000-10,000 B.Th.U. It can absorb water up to about 40 per cent. On air drying, it tends to crumble and become powder. The volatile matter consists mainly of nitrogen and also of hydrogen and oxygen. The ash contains, inter alia, iron and magnesium. Lignite is soluble in caustic soda. It is thermostable at high temperatures prevailing at depths below the earth's surface.

COMMON USES

1. Domestic and Industrial Fuel

Due to its thermal value, lignite may serve as a useful (though inferior to coal) fuel for both domestic and industrial use. It is particularly useful in cement and other medium and small scale industries in and around lignite-bearing areas. Besides somewhat low thermal value, the other problem is its tendency to crumble into powder on heating. For this reason, lignite is made into formed fuel such as pellets by using some binding substance. Now-a-days binderless briquettes are manufactured by crushing the raw lignite to 4.6 mm size, drying either in steam or flue gas, cooling to about 40 °C and then finally briquetting under a high pressure of 10-12 tonnes/sq.inch. The briquettes are carbonized at 600-650 °C.

2. Thermal Power

Lignite is used both as an independent fuel and as a blend with coal and for generation of steam to move turbines and produce power in the same way as coal (see also the chapter 'Coal'). The lower thermal value is compensated by the advantage of low ash-content (and consequently less problem of fly-ash disposal).

3. Fertilizer and Chemicals

Lignite contains fairly high quantities of volatile matter. On carbonization volatile matter is expelled and being condensed, settles down as tar and ammonia layers. By fractional distillation, the ammonia is recovered. From this ammonia, nitrogenous fertilizers like $(NH_4)_2SO_4$ is manufactured. By fractional distillation of the tar, various other chemical derivatives of volatile matter can also be recovered. (See also the chapter 'Coal').

4. Biofertilizer

The decomposed vegetable matter and the water-holding capacity make lignite suitable for culturing of some bacteria beneficial to plants. In this use it serves as an inferior substitute of peat. (See also the chapter 'Peat').

5. Oil Well Driling

Causticized lignite (also called sodium lignite) is used as an additive to drilling mud for reducing it rheological properties and fluid loss. Causticized lignite is more effective than common mud chemicals in deep driling when thc temperature may exceed even 200°C (whereas the common mud chemicals start becoming less and less effective at temperatures above 120°C). For this use lignite containing 15% (maximum) moisture is ground to 70 mesh powder, mixed with 2.5% caustic soda in ratio of 2-3 (lignite):1 (caustic soda) and digested by stirring. The prepared solution is then aged for 24 hours at 32°C, dried, cooled and crushed to required size.

Causticized lignite acts as an emulsifying agent in oil-water emulsion. An emulsion is a dispersion of liquid in another immiscible liquid. Perfect emulsion of two liquids is not possible unless an emulsifying agent is present. The suspended droplets of the dispersed liquid adhere to the emulsifying agent, and thus they are prevented from coalescing. In case of the oil-water emulsion, as prevalent in drilling mud, causticized lignite powders are preferentially wetted by oil. Subsequently, that oil can be recovered.

6. Lignite-based Met-coke

For manufacturing coke from lignite, crushed lignite is first dried, then carbonized and finally calcined at successively higher temperatures in a three-bed fluidized system. The calcinate is then mixed with some suitable binder and briquetted. This green briquette is first cured on a moving grate in an oxygen-containing atmosphere and then devolatilized to yield coke. However, the technology of manufacturing lignite-based coke requires costly investment. About 1.5-2.3 tonnes of lignite can yield one tonne of coke.

This met-coke can be an economic source of fuel for the cement, soda ash, fertilizer, textile and steel plants located in lignite-bearing areas where coal is either not available or, if available, is very costly.

7. Reductant

In Western Australia, in the beginning of 21st century, intensive research has been undertaken to gauge the practicality of using lignite as a reductant for the processing of iron ore through to metal. The laboratory scale tests have shown encouraging results.

8. Pig Feed

According to the finding of research carried out in the Pig Institute in Wrocklaw, Poland, feeding lignite to pigs as a dietary supplement is good for their health. Pigs fed on brown coal were found to be fatter, happier, healthier and less stressed than ones fed on standard chemical additions. Experiments showed that lignite provided the pigs with a source of magnesium and iron. It increased their haemoglobin levels by around 6%, thereby substantially reducing cases of anaemia. Organic acids in the lignite absorbed toxins from the digestive system and improved bacterial flow preventing diarrhoea.

Subsequently, the Wrocklaw Institute of Petroleum and Coal has reported the possibility of converting the excreta of lignite-consuming pigs into cheap and effective fuel, which may well turn out to be a source of renewable energy. Pigs that eat lignite, have very solid dark excrement which can be dried and burned.

7

CHAPTER

MONAZITE AND THORIUM

The concentration of thorium in the earth's crust has been estimated to range from 10 to 20 ppm. Thorium is the major constituent of thorite (silicate of thorium, uranium, iron, manganese, copper, magnesium, lead, tin, aluminium, sodium and potassium) and thorianite (oxide of thorium, uranium and other rare earth metals). But by far the most important commercial source of thorium is monazite. It may be a constituent of pegmatites, granites and gneisses; but commercial deposits of monazite occur in the form of placer. Monazite sands are concentrated in Brazil, India, Sri Lanka, Indonesia, Australia, South Africa, Malaysia, Canada, Greenland and USA. Lesser quantities of thorium-bearing monazite reserves occur in vein deposits and carbonatites. Monazite is a complex phosphate of cerium (Ce), lanthanum (La), praseodymium (Pr), neodymium (Nd), samarium (Sm) with small amounts of the rare earth elements of the terbium and yttrium groups, thorium and occasionally uranium.

HISTORY

Thorium was discovered in 1828 by a Swedish chemist Berzelius when he isolated a new element from a mineral occurring in Norway. The name was taken after 'Thor', the warrior god of thunder of the Nordic race.

The monazite sand deposit of Kerala in India was discovered in 1909. Coir workers of that region used to rub their hands in sand to get a grip of the coir. Some sand used to stick to the wet coir. Eventually, some sand found its way to Germany along with the exported coir. One day, in 1909, C. W. Schomberg, a chemist, by chance, stumbled upon this glistening sand. He could identify that the sand contained monazite, an important material for gas light mantle. He came to India and located the deposits. He established a separation plant in 1911, and in that year exploitation also commenced. The British took over it during World War-I (1914-18).

In 1946, interest in thorium as a possible nuclear energy source began developing, and the Government of India stopped exporting monazite after independence (1947). In 1950, a new undertaking, namely the Indian Rare Eaths Ltd. (IREL) was created, and a monazite processing plant was commissioned in 1952 in Alwaye, Kerala.

The production statistics of Indian manazite for the period 1911-1949 are mostly based on old reports. The available data are as follows:

Year/period	Average annual production
1911-1915	1,016 tonnes
1918	2,152 tonnes
1922-1931	176 tonnes
1932	664 tonnes
1934	1,025 tonnes
1938	5,305 tonnes
1949	5,080 tonnes

Since 1949, statistics related to monazite have stopped being published for strategic reasons.

Although both thorium-fuelled burner and breeder reactors were developed in the 1960s and 1970s, but fell behind thereafter due to lack of enthusiasm about nuclear power in general, and a more focused development of uranium-fuelled nuclear power technologies.

CRITERIA OF USE

The physical, chemical and mechanical properties that hold the key to practical use of monazite and thorium are as follows:

1. Chemical Composition of Monazite

Theoretically, the formula for monazite should be (Ce, La) PO_4. But thorium and yttrium earths tend to substitute for cerium partially, and the most accepted formula is (Ce, La, Y, Th) PO_4. In addition, presence of Fe, Al, Ca, Mg, Si, Ti, Zr and sometimes U is common. Consequently, the ThO_2-content in monazite varies from place to place. It generally varies from about 1 to 11 per cent. But, monazite itself does not occur in nature as such. While its content in some hydrothermal veins has been reported to be as high as 70%, the most common source—namely placer sand—may contain as low as 1 per cent monazite. In placer sand, it is associated with different heavy minerals like ilmenite, rutile, garnet, zircon, sillimanite, etc. Consequently, extraction of the useful thorium from monazite involves several steps.

(*a*) The first step is preliminary concentration of the monazite from the placer sand. This is done by separating out lighter sands through tabling or sluicing. The concentrate may contain 20-60% monazite.

(*b*) The second step is to concentrate it further to marketable grade containing at least 95% monazite. This is achieved through electromagnetic separation. The monazite sand contains the following minerals in order of decreasing magnetic permeability:

- Magnetite (strongest magnetic permeability)
- Ilmenite and haematite
- Garnet, platinum, epidote, apatite, olivine, tourmaline
- Monazite
- Zircon, rutile, gold (nonmagnetic)

This difference in the degree of magnetic permeability is made use of in separating these groups of minerals one by one by deploying progressively higher intensity magnets. Monazite, being weakly magnetic, is separated at the last stage by deploying very high intensity magnets.

(*c*) After the physical processes of concentration, the next step is chemical extraction of thorium concentrate from the monazite concentrate. For this purpose, monazite is digested in either H_2SO_4 or NaOH. A mixture of either sulphates or hydroxides of thorium, rare earths and uranium is extracted. Thorium content in this mixture may be up to 50 per cent.

(*d*) The next step is purification of the thorium concentrate. In the usable thorium metal, even 1 ppm of any neutron-absorbing rare earth element like gadolinium, samarium, dysprosium, etc., is not acceptable. There are three principal methods of purification: (*i*) solvent extraction, (*ii*) fractional crystallization and (*iii*) selective leaching.

In the solvent extraction method, organic solvents like methyl isobutyl ketone, nitromethane, naphthyl methyl ether, diethyl oxalate, etc., may be used. These solvents have the propensity to dissolve thorium more than the impurities.

In the fractional crystallization method, thorium hydroxide along with impurities is dissolved in sulphuric acid. Then thorium sulphate, being less soluble than cerium sulphate, is fractionally crystallized. The thorium sulphate is then reconverted to thorium hydroxide by treating with ammonia. Thorium hydroxide can then be treated further with nitric acid to yield thorium nitrate tetrahydrate.

In the selective leaching process, mixed oxalates of thorium and rare earths are treated with sodium carbonate and then filtered. Thorium carbonate (which is soluble) is formed, whereas the carbonates of rare earths are left behind as residue. By treating the thorium carbonate solution with caustic soda, purified thorium hydroxide is precipitated, which can then be further purified by dissolving in hydrochloric acid and selective precipitation with sulphuric acid.

(*e*) The next step is to convert the purified thorium concentrate—usually in the form of sulphate or nitrate—into thorium oxide (ThO_2) or thorium tetrafluoride (ThF_4) or thorium tetrachloride ($ThCl_4$), which are the most common reducible salts of thorium. ThO_2 may be produced by first forming thorium oxalate and then calcining the oxalate. ThF_4 may be produced by treating either the oxide or the nitrate tetrahydrate of thorium with HF-acid. $ThCl_4$ may be produced by treating ThO_2 with carbon tetrachloride or chlorine.

(*f*) The next step is to reduce the thorium salts to metallic thorium. This is achieved by calcium reduction of ThF_4, or by magnesium reduction of $ThCl_4$ or by calcium reduction of ThO_2 or by electrolytic methods, out of which the first one appears to be the most common in commercial production of thorium.

(*g*) Finally, the thorium metal is melted and cast in desirable shapes.

(*h*) During manufacturing of $Th(NO_3)_4$, an isotope—mesothorium—is also recovered as a byproduct.

It is the thorium and mesothorium, contained in monazite, that make it useful in various applications.

2. Radioactivity

This phenomenon is explained in the chapter 'Uranium ore and metal'. Thorium is poorly radioactive. Its half-life is 13.9 billion years (cf. 4.5 billion years of uranium). Its radioactive decay is through emission of alpha rays. Its final stable decay product is lead-208. However, in between, 10 intermediate isotopes are formed which have shorter half-lives ranging from 0.3 microseconds to 6.7 years. Some of these isotopes emit beta rays and some alpha rays. Only one important isotope namely mesothorium (atomic number 88, atomic weight 228), which is intensely radioactive (half life 6.7 years), emits both beta and gamma radiation (0.09 MeV). This isotope is usually found in association with thorium in monazite, and can be recovered as a byproduct. Its oxide is also intensely radioactive.

3. Nuclear Fission

This phenomenon is explained in the chapter 'Uranium ore and metal'. Th_{232} is not itself fissionable. On being bombarded with a neutron, its atom does not split, but the neutron is captured; and this initiates a series of changes within the atom through beta radiation, yielding finally a new element U_{233}:

$$_{90}Th_{232} + 1 \text{ neutron} \rightarrow {}_{90}Th_{233}$$

$$_{90}Th_{233} - 1 \text{ beta} \rightarrow {}_{91}Pa_{233}$$

$$_{91}Pa_{233} - 1 \text{ beta} \rightarrow {}_{92}U_{233}$$

Note: Th = Thorium
Pa = Protactinium
U = Uranium

In this process, the total mass of the neucleons (neutron + proton) increases by one (from 232 to 233) due to the additional neutron initially captured. However, the number of protons also increases by two from 90 to 92 on account of transformation of two neutrons into two positively charged protons as soon as two negatively charged beta particles are emitted therefrom. Now, this U_{233} atom is fissionable and it can sustain chain reaction. On being hit by a neutron, it splits and emits 2-3 neutrons. One of these released neutrons can hit another atom of U_{233} and start chain fission, while the other ones may strike surrounding Th_{232} atoms, and more number of U_{233} atoms can be bred. In this way, Th_{232} can take part in chain fission.

4. Ductility

This is the opposite of tensile strength and it signifies the ease with which a metal can be drawn into a wire without breaking. Pure thorium has high ductility. Even small amounts of impurities—particularly carbon, and also nitrogen, oxygen and indium—markedly increase its tensile strength and reduce ductility.

5. Melting Point

Values of melting point of thorium metal ranges widely from about 1500°C to about 1800°C depending on traces of impurities. However, ThO_2 is one of the most refractory substances known, its melting point being of the order of 3000°C.

6. Chemical Reactivity

Thorium is a highly reactive metal with electropositive characteristic. It forms binary compounds with all the nonmetallic elements, except the rare gases. It has a particular affinity for oxygen with which it combines easily giving off large quantities of free energy.

7. Adsorption

Thorium metal and oxide are capable of adsorbing many vapours and gases without chemically combining with them. Such adsorbed gases become available for chemical reaction with other substances, while thorium itself remains unaffected.

8. Luminosity

ThO_2 , on heating, burns with a bluish light. Presence of about 1% cerium renders the flame whiter and brighter.

9. Alloying Behaviour

An alloy is a substance composed of two or more metals intimately mixed and united, usually being fused together and dissolving in each other when molten. The alloying substances form a solid solution, and for this, their crystal structure and electrochemical properties should be similar.

Further, the substitutional and interstitial size-fit criteria are important for the atoms of the alloying elements to form a solid solution. For a good substitutional size-fit, the atomic diameters of the solute and solvent atoms must not differ by more than 14-15%, and for good interstitial size-fit, the ratio of the radius of the smaller atom to that of the larger atom should be less than 0.59. Thorium can form alloys with several metals.

10. Optical Properties

Fresh and pure thorium metal is silver coloured. But the metal being highly reactive, it tends to form hydride on long exposure to moisture, or oxide on being heated in air, and consequently, colour changes. However its most characteristic properties are low dispersion, low yet consistent emissivity and high refractive index.

Dispersion is the rate of change of refractive index with change in wavelength of the incident light, and is expressed with reference to some wavelength. Low dispersion signifies that when a ray of light is incident on thorium surface, it will not appreciably break up, after refraction, into its different colour components having different wave lengths. In other words, the chromatic composition of the incident and refracted rays will be more or less similar.

Emissivity is a measure of the energy (heat or some other form) appearing within a substance due to absorption of incident light. A perfectly black substance absorbs all the incident light, converts it into some radiation energy and may emit the same. The emissivity of such a substance is reckoned as '1', and this serves as the reference standard. Since all the nonblack objects absorb less light than a black one, their emissivity is always less than 1; and lower the emissivity, more will be the amount of light transmitted (zero emissivity means that absolutely no light is absorbed). Thorium has fairly low emissivity (of the order of 0.35-0.40). also, a striking characteristic of thorium is that this emissivity value does not vary widely with change in wavelength of the incident light, or in temperature of the metal, or in state of the metal (i.e., solid or liquid). This low emissivity value signifies that (*i*) thorium does not absorb much light, and (*ii*) the degree of absorption is more or less the same for different colours of light, i.e., there is not much differential absorption of the different colours.

11. Electron Emission

Electrons present in the crystal lattices on the surface of a metal can be liberated by the addition of energy, in different forms such as light rays (photoelectric emission), heat (thermionic emission) or electric current (field emission), etc. The external energy agitates the atoms of the metal; as a result high-energy electrons overcome the intra-atomic forces, break out from the surface of the metal and escape. This is the principle of electron emission. Thorium requires comparatively little energy to produce high electron emission.

COMMON USES

In some of the applications, thorium metal is used, in others, its compounds (particularly ThO_2) and alloys. The thorium-related common uses are:

1. Nuclear energy
2. Alloys
3. Chemical compounds
4. Catalytic agent
5. Deoxidant
6. Electron tubes
7. Special refractories
8. Gas mantle
9. Lamp filaments
10. Radiotherapy
11. Optical glass
12. Welding
13. Paint

These are discussed as follows:

1. Nuclear Energy

As has been said, Th_{232} is not itself fissionable and so it cannot by itself produce nuclear energy (see also the chapter 'Uranium ore and metal'). However, on being hit by a neutron, it is capable of absorbing it, and of being finally transformed into U_{233}, which is fissionable by both thermal and fast neutrons. Since U_{233} isotope is not naturally occurring, it requires Th_{232} for its production, and thorium metal or oxide is indirectly useful in the generation of nuclear energy. There are mainly two principles on which the mechanism of thorium-based power generation may work.

(*a*) Thorium is first used in a thermal reactor in conjunction with natural or enriched uranium which contains mostly U_{238} and a little U_{235}. The latter is a natural fissile substance capable of emitting fast neutrons. The neutrons in excess of those required to sustain chain reaction of U_{235}, hit the surrounding U_{238} and Th_{232} atoms. As a result Pu_{239} and U_{233} are produced. These two fissionable substances can be chemically separated, and the U_{233} can be used in a thermal reactor.

(*b*) During fission, Pu_{239} emits fast neutrons. So, the Pu_{239} obtained from a thermal reactor, may be used in conjunction with Th_{232} in a fast breeder reactor. The neutrons emitted from Pu_{239} can hit the Th_{232} atoms and eventually U_{233} may be produced. These U_{233} atoms then can take over the breeding process. Some of the neutrons emitted from them will hit other U_{233} atoms and sustain a chain reaction, while the excess neutrons will hit the unconverted Th_{232} atoms and yield more U_{233} atoms. Thus the Th_{232} may serve to sustain a breeding process, by which more U_{233} atoms will be bred than consumed. In this method also, both thorium metal and oxide may be useful.

In either process, the fission eventually results in loss of some mass of the original atom, and energy in the form of heat is released in accordance with Einstein's equation $E = mc^2$ (where 'E' is energy, 'm' is mass loss and 'c' is speed of light, i.e., 300000 km/sec). The heat thus produced is used to convert water into superheated steam for moving turbines and generating electricity as in the case of thermal power generation (see the chapter 'Coal').

2. Alloys

Thorium alloys readily with iron, cobalt, nickel, copper, gold, silver, platinum, molybdenum, tungsten, tantalum, zinc, bismuth, calcium, lead, mercury, sodium, beryllium, silicon, cerium, chromium, zirconium, lithium, magnesium, antimony, tin, thallium, uranium, indium, titanium and niobium. Pure thorium metal is worthless for structural engineering because of its low tensile strength, low elastic modulus and poor resistance to corrosion. Alloying of thorium seeks to improve certain mechanical properties like strength, welding characteristics, etc.

3. Chemical Compounds

Thorium, being highly reactive chemically, combines with non-metallic elements and metals, and yields useful compounds and salts as follows:

(*i*) Hydride–ThH_2, ThH_3, ThH_4
(*ii*) Oxide–ThO_2, ThO
(*iii*) Hydroxide–$Th(OH)_4$
(*iv*) Peroxide–$Th_2O_7 \cdot 4H_2O$
(*v*) Nitride–Th_2N_3, ThN
(*vi*) Nitrate–$ThO(NO_3)_2 \cdot H_2O$, $ThO(NO_3)_2 \cdot 5H_2O$, $ThO(NO_3)_4 \cdot 4H_2O$
(*vii*) Chloride–$ThOCl_2$, $ThCl_4 \cdot 9H_2O$, $Th(OH)Cl_3 \cdot 11H_2O$, $ThOCl_2 \cdot 3H_2O$, $ThOCl_2 \cdot 5H_2O$
(*viii*) Fluoride–$ThF_4 \cdot 8H_2O$, $Th(OH)F_3 \cdot H_2O$, $ThOF_2$
(*ix*) Bromide–$Th(OH)Br_3 \cdot H_2O$, $ThOBr_2$
(*x*) Iodide–ThI_4, ThI_3, $Th(OH)I_3 \cdot 10H_2O$
(*xi*) Iodate–$4Th(IO_3)_4 \cdot KIO_3 \cdot 18H_2O$
(*xii*) Sulphide–ThS_2, Th_2S_3
(*xiii*) Oxysulphide–ThOS
(*xiv*) Carbide–ThC, Th_2C_3
(*xv*) Carbonate–$ThOCO_3 \cdot 8H_2O$
(*xvi*) Oxalate–$Th(C_2O_4)_2 \cdot 6H_2O$
(*xvii*) Phosphide–Th_3P_4
(*xviii*) Phosphate–$Th_3(PO_4)_4$, $Th(HPO_4)_2 \cdot H_2O$, $Th(HPO_4)(H_2PO_4)_2 \cdot 2H_2O$
(*xix*) Chromate–$ThO_2 \cdot 2CrO_3 \cdot 3H_2O$

Some of the basic compounds of thorium form double salts with compounds, such as double salts of thorium fluorides with sodium and potassium, double salts of thorium chloride and ammonia, etc., and also various useful thorium salts like sulphates, molybdates, etc.

4. Catalytic Agent

The property of adsorption of many vapours and gases makes thorium useful as a catalytic agent in certain chemical reactions such as in oxidation of SO_2 to SO_3, in production of water gas (a mixture of CO and H_2) by passing steam over hot coal, in production of HNO_3 from NH_3, etc. A mixture of thorium and cobalt in the ratio of 1:20 has been found to be a possible catalytic agent in the synthesis of hydrocarbons.

5. Deoxidant

Strong affinity of thorium for oxygen makes it useful in the reduction of metals like molybdenum and iron.

6. Electron Tubes

If thorium electrodes are placed in a discharge tube containing impure inert gas, the metal rapidly consumes the oxygen and nitrogen present and keeps the inert gas pure. Besides, the high electron emission of thorium electrodes offers the advantage of lower starting potential. In this application, the strong chemical affinity of thorium towards oxygen and nitrogen (remaining at the same time nonreactive to inert gases) and the property of electron emissivity are the key.

7. Special Refractories

The high melting point of 3000°C makes ThO_2 useful in construction of special refractories such as crucibles for laboratory melting of vanadium, titanium, etc.

8. Gas Mantle

Before the advent of electricity, this used to be the most important use of monazite. Thorium nitrate or oxide obtained from monazite can be used in the manufacturing of incandescent gas mantles on account of its brilliant luminosity on heating. Thoria itself emits a bluish light, whereas addition of 1-2% cerium oxide makes the light whiter and brighter. Even long after the advent of electric light, some special types of kerosene lamp and gasoline lantern with thoria-impregnated gas mantles have been used by armed forces in areas remote from power lines.

9. Lamp Filaments

In this application, an alloy of thorium with tungsten is used. Pure tungsten filament, after short use, tends to crystallize, becomes hard and brittle, and eventually breaks. Addition of 0.8-1.2% thorium inhibits the crystal growth, controls grain size, increases ductility and prolongs the lives of filaments considerably. The ductility of thorium metal is the key criterion for this purpose. However, in this use, carbon, oxygen, nitrogen and indium are highly deleterious, because these elements tend to increase the tensile strength and to correspondingly decrease the ductility of thorium. Instead of thorium metal, incorporation of its oxide in tungsten can also serve the same purpose.

10. Radiotherapy

Thorium as such is very weakly radioactive, but the intermediate decay product mesothorium, being intensely radioactive, has been valued in radiotherapy. However, on account of its intense radioactivity and consequent high degree of radiological toxicity, it is usually reserved only for emergencies as the last resort.

11. Optical Glass

An optical glass differs from ordinary glass in its freedom from bubbles and chemical inhomogeneity—the factors responsible for development of regions of variable refractivity in ordinary glass. Pure thorium metal is added to optical glass for use in photographic lenses. Due to its characteristic optical properties of high refractive index, low dispersion and low but consistent emissivity, thorium increases the refractive index of the lens, lower the dispersion of light, ensures more or less the same chromatic composition in the transmitted light as in the incident light, and at the same time, it does not absorb much of the incident rays of light.

12. Welding

The mechanism of welding is based on electron emission. In this, the electron discharge takes place in the form of an arc. When electricity is passed though two electrodes (cathode and anode) in contact with each other, and then the contact is broken by moving them a little away, the resistance and consequently the potential, increases so much that the tips of the electrodes begin to glow. The temperature at the tips increases rapidly, and electron emission takes place. The high energy electrons associated with the temperature ionizes the air in the gap between the electrodes. This ionized air becomes an electrical conductor and current flows from one electrode to the other. This is the mechanism of arc discharge. The temperature of the arc may be of the order of thousands of degrees (20000-50000 °C). If the broken pieces of a metal are placed in the arc, then they will fuse and join together, and this process is known as welding. The electrodes are called welding rods. If the metal to be welded itself is an electrical conductor, then it may serve as the second electrode, and only one welding rod will be required. If the welding rod is made up of a fusible metal which can mix with the fused welded metal and thus strengthen the weld, then the welding rod is called 'consumable' and it requires replenishment. If the welding rod remains in tact and only the welded metal fuses to form the weld, then the rod is called 'nonconsumable'.

Tungsten is a common non-consumable welding rod material. But it requires high starting potential. On the other hand, addition of thorium to the tungsten reduces the starting potential for electron emission and consequently engenders instant arc discharge and instant arc stability.

13. Paint

The intermediate radioactive decay product of thorium, namely mesothorium, is highly radioactive. Its half life (6.7 years) is much shorter than that of radium (1600 years). Since ThO_2 invariably contains some mesothorium, it can be used as a substitute of radium in luminous paints.

14. Other Uses

The property of high electron emission makes thorium useful in electrodes of mercury arc lamps, in low pressure cathode lamps, for anode coating of radio bulbs and in photoelectric cells meant for measuring intensities of X-ray and ultraviolet light.

8

CHAPTER

PEAT

Peat signifies a preliminary stage in the process of coalification of vegetable matter. The decayed vegetable matter of the geological past has already resulted in the formation of higher members in the series namely lignite, bituminous coal, etc., and peat cannot be expected to be associated with such minerals. Peat can only be expected in comparatively recent formations where partially decayed vegetable matter (humus) had accumulated in a water saturated environment in the absence of oxygen, in bogs, swamps, marshes, which have practically no drainage; and has subsequently been buried. Although, it is not an energy mineral in the strict sense, it is used as a sort of fuel in one or two commercial uses.

CRITERIA OF USE

Peat is a spongy material, and in raw state it contains about 80% moisture, which on air drying may come down to about 20 per cent. But this behaviour signifies a high capacity to absorb and hold water. Besides moisture, peat contains partially decomposed vegetable matter, and, on account of this, possesses some thermal value also.

COMMON USES

1. Scotch Whisky

This is an important commercial application of peat in which its thermal value is made use of. In the manufacturing of Scotch whisky, barley is converted to malt by allowing it to sprout. Then it is dried in a kiln over a peat fire. Malt absorbs some of the smoke aroma which is carried over later with the spirit distilled from it. The peat smoke imparts the special flavour.

2. Biofertilizer

The partially decomposed organic matter provides a suitable setting for certain bacteria, which are beneficial for plant growth to thrive. Peat serves as a carrier for such bio-fertilizers. Besides, the water-holding ability of peat also adds to the suitability of peat in this application.

3. Soil Amendment

Powdered peat may be directly applied to soil as a manure to improve the water absorption and water retention in the soil.

4. Fuel

The air-dried peat can be used as a slow burning fuel. It has been used in some areas as a fuel for lime and brick kilns.

5. Preservative

The water-absorbing property of peat makes it suitable for use in packing material for preservation of fruits, vegetable, etc., during long transportation by sea, railways, etc. In this application, it is particularly useful in dust form.

6. Surgery

As it holds water well, it is used for surgical dressings.

SUBSTITUTION

In U.S.A. a compost processed from urban garbage has been reported to be suitable as a substitute of peat in its application in biofertilizer and soil amendment. Its trade name is 'agrisoil'.

9

CHAPTER

ANTHRACITE

Anthracite is the final stage in the process of coalification. It was first known to be used in 1755 by gunsmiths in Pennsylvania, USA.

CRITERIA OF USE

It is hard and brittle. It contains a high percentage of fixed carbon (up to 86% on dry basis), low ash and low volatile matter. Samples of anthracite mined in Vietnam contained about 90% fixed carbon, 7% ash, 3% volatile matter and 0.8% sulphur. Its thermal value is high (over 13000 B.Th.U per pound on dry basis).

COMMON USES

Its high thermal value makes it very valuable as a fuel, and it can be used in iron making, in thermal power generation, and in domestic and industrial heating in the same way as coal (see also the chapter 'Coal'). Being very low in volatile matter content, it is difficult to catch fire initially, but once ignited, it burns with a smokeless short blue flame. The low ash content is another advantage. In order to overcome the problem of initial ignition, it is sometimes blended with bituminous coal and made into briquettes. Due to the low volatile matter content, anthracite cannot be carbonized to make coke (see also the chapter 'Coal'). However, when a blend of coking bituminous coal and anthracite is carbonized, the resultant coke will have higher thermal value than that produced from bituminous coal alone. The anthracite mined in Vietnam is used for iron ore sintering, lime-burning and calcium carbide manufacturing.

"Knowledge is the only instrument of production with no diminishing return".

— J. M. Clark

10

CHAPTER

BIBLIOGRAPHY

In course of writing this book, published literature, websites of United Nations Economic Co-operation for Europe (UNECE), proceedings of many seminars, personal correspondence and interaction with industries and institutes, press reports, etc., served as the main plank for sourcing of information. The published literature included various editions of the erstwhile United States Bureau of Mines (USBM) publication Mineral Facts & Problems from 1952 onwards, issues of the Indian Bureau of Mines (IBM) publication Indian Mineral Year Book from 1964 onwards, the quarterly issues of E-Coal published by the World Coal Institute (London), old and recent fortnightly issues of FIMI News brought out by the Federation of Indian Mineral Industries, monthly issues of the 'Mining Engineers' Journal brought out by the Mining Engineers' Association of India, papers and reports published in various Indian and international technical journals, Encyclopedia, and some textbooks. Some important books and papers recommended for further reading are listed below by author's name in alphabetical order.

1. Bateman Alan M. (1962): Economic Mineral Deposits; 2nd edition, Asia Publishing House, 916 p.
2. Bone, William A. (1918): Coal and its Scientific Uses; Lngman, Green & Co, London, 491 p.
3. Chandra D., Roy P., Mishra A.K., Chakraborti J.N., Prasad N.K. & Chaudhuri S.G.: Removal of Sulphur from coal by Thiobacillus Ferroxidans and by Mixed Acidophilic Bacteria Present in Coal; Fuel, London, 1980, Vol. 59, April, p. 249-252.
4. Chandra D., Chaudhuri S.S.G. & Ghosh: Distribution of Sulphur in Coal Seams with Special Reference to the Tertiary Coals of North-Eastern India; Fuel, London, 1980, Vol. 59, May, p. 357-359.
5. Chandra D., Roy P., Mishra A.K., Chakraborti J.N. & Sengupta B: Microbial Removal of Coal; Fuel, London, Vol. 58, July, p. 549-550.
6. Coggin Brown J. & Dey A.K. (1955): India's Mineral Wealth; 3rd edition, Oxford University Press, 761 p.
7. Cuthbert F.L.: Thorium production technology; Addison-Wesley Publishing Co., USA, 1958, 303 p.
8. Dunstan A.E. (1920): The Petroleum Industry: Institute of Petroleum Technologists, London, 346 p.

9. Gitthus J.H.: Uranium; Butterworths, London, 1963, 623 p.

10. Holden A.N.: Physical Metallurgy of Uranium; Addison-Wesley Publishing Co., USA, 1958, 262 p.

11. Jevons, Stanley H. (1915): The British Coal Trade; Kegan Paul, Trench, Trubner & Co. Ltd., London, 876 p.

12. Lowry H.H. (1963): Chemistry of Coal Utilization (supplementary volume); John Wiley & Sons, 1142 p.

13. Redwood B. (1926): Petroleum; Charles Griffin Co. Ltd., London; Vols I, II & III, 1353 p.

14. Sell G. & Dossett H.A. (1958): Handbook of Petroleum Industry: George Newnes Ltd., London, 253 p.

15. Singh M.P.: Trace Elements in Coal; Minetech, CMPDI, Vol. 12, no.1, Jan-Feb, 1991, p. 16-46.

16. Wilson Philip J. & Wells Joseph H. (1950): Coal, Coke and Coal Chemicals, McGraw-Hill, 509 p.

GLOSSARY

Alloy: An alloy is a substance composed of two or more metals intimately mixed and united, usually being fused together and dissolving in each other when molten. The alloying substances form a solid solution, and for this, their crystal structure and electrochemical properties should be similar.

Carbon credit: A carbon credit is a unit that measures a specific amount of reduction of green house gases (GHG). These credits are generally represented as a GHG reduction equivalent to a tonne of carbon dioxide or carbon or methane.

Cenosphere: Cenosphere is a silicate glass filled with nitrogen and CO_2, and it is produced due to conversion of a portion of the fly ash during the combustion process.

Cracking: In 'cracking', molecules are broken down under high temperature (with or without a catalyst) into smaller units, and a new type of hydrocarbon namely olefin is produced. By cracking, light gases, petroleum coke, fuel oil etc., can also be produced.

Dielectric strength: Dielectric strength is a measure of the electrical insulation, and is the voltage that an insulating material can withstand before break-down. It is expressed in terms of specific resistance.

Dispersion: Dispersion is the rate of change of refractive index with change in wavelength of the incident light, and is expressed with reference to some wavelength.

Electron emission: Electrons present in the crystal lattices on the surface of a metal can be liberated by the addition of energy, in different forms such as light rays (photoelectric emission), heat (thermionic emission) or electric current (field emission) etc. The external energy agitates the atoms of the metal; as a result high-energy electrons overcome the intra-atomic forces, break out from the surface of the metal and escape. This is the principle of electron emission.

Emissivity: Emissivity is a measure of the energy (heat or some other form) appearing within a substance due to absorption of incident light.

Emulsion: An emulsion is a dispersion of liquid in another immiscible liquid.

Fission: The ability of an atom to split due to collision with a free neutron is called 'fission'.

Gross calorific value: Gross calorific value is the total amount of heat obtainable by the combustion of a given coal. Its units are kilocalorie and British Thermal Unit or BTU or

B.Th.U. Kilocalorie denotes the number of kilograms of water which may be heated through 1°C, in the neighbourhood of 15°C, by the complete combustion of 1kg of coal. BTU denotes the number of pounds of water which may be heated through 1°F, in the neighbourhood of 60°F, by the complete combustion of 1lb. of coal. In either of these cases, the conditions are: (*i*) coal is dried at 105°C until its weight becomes constant, (*ii*) whole of heat is transferred without loss to the water, and (*iii*) the products leave the system at the atmospheric temperature and pressure.

Half-life: The period in which the number of atoms of a radioactive substance decreases to one half its original value (with proportional increase in the mass of lead produced) is called 'half-life'.

Isomerization: Isomerization is the process of producing a similar but new substance by rearrangement of atoms within the hydrocarbon molecules of the original substance.

Net calorific value: Net calorific value is the gross calorific value minus the heat liberated by the condensation of the steam produced on combustion and the subsequent cooling of this condensed steam to water down to atmospheric temperature (15°C or 60°F).

Octane number: Octane number is a measure of 'anti-knock' value of a motor fuel i.e. the ability to resist the knock or sound produced due to its sudden and violent combustion in a spark ignition engine. For this measurement, a standard scale has been devised by assigning the value zero to heptane (C_7H_{16}) which has very poor knock resistance, and 100 to octane (C_8H_{18}) having a very high knock resistance. Octane number is the percentage of this isomer of octane in its mixture with heptane.

Radioactivity: Radioactivity is the spontaneous disintegration of certain heavy elements accompanied by the emission of high energy radiation, which consists of three kinds of rays: 'alpha particles', 'beta particles' and 'gamma rays'.

Reforming: Reforming is a special type of cracking in which a heavy low-octane naptha is processed for octane improvement rather than volatility change.

Sialon ceramics: It is an advance material comprising a mixture of silicon, aluminium, oxygen, and nitrogen (i.e. Si-Al-O-N). Sialon is suitable for applications requiring high mechanical strength at elevated temperatures, high specific strength (for weight saving without sacrificing strength), high hardness and toughness, low coefficient of friction and good thermal shock resistance.

Viscosity: Viscosity is that property of a liquid which is a measure of its internal resistance to motion and which is manifested by its resistance to flow.

INDEX

A

Acetylenes 55
Acid rain 31
Acrylic fiber 74,77
Activated carbon 24
Adsorption 104
Agricultural waste energy 44, 51
Agrisoil 110
Alkylation 62
Alkynes 55
Alloy 104
Alloy system 88, 89
Alpha particles 87
Alpha phase of uranium 88
Ammonia 12, 18
Animal energy 72
Anthracene oil 12, 18
Anthracite of Vietnam 111
Antiknock value 62
Apsara 86
Aromatics 52, 55
Ash fusion temperature 11, 16
Asphalt 52, 67
Asphaltic bitumen 67
Asphaltic rocks 71
Assam Gas Company Ltd. (AGCL) 76
Assam Oil Company Ltd. 53, 54
Assam Oil Syndicate Ltd. 53
Assam Railway and Trading Co. 53
Atom bomb 91, 92
Atomic Energy Commission 86
Aviation Turbine Fuel (ATF) 58, 65

B

Babnizu mine of Iran 27
Bacteria energy 44, 51
Bacterial demineralization 29
Bacterial leaching of sulphur 26
Badarpur Oil Syndicate 54
Barrel technology 13, 14
Beehive 9, 21
Bell laboratory 46
Benz Carl 53
Benzene 12, 18
Benzenoid 55
Benzole 12, 18
Bequerel, Antionne Marie 85
Berzelius 100
Beta particles 87
Beta phase of uranium 88
Bhabha, Dr Homi Jahangir 86
Bio-fertilizer 43, 109, 110

Biogas 41
Bio-gasification of mine sludge 28
Biomass energy 44, 51
Bitumen 58, 67
Blue water gas 19
Bomb calorimeter 8
Brick burning 21
Briquette 26, 97, 98
Bruce 53
Bunsen burner 19
Burma Oil Company Ltd. 54

C

Caking index 9, 10
Caking property 9
Calcium carbide 24
Calico printing 93
Candle making 67
Candle power 69
Carbolic acid 12,18
Cresol 18
Carbon black 69, 78
Carbon-carbon composite 18
Carbon credit 40
Carbon dioxide as an economic Commodity 30
Carbon fiber 18
Carbon sequestration 30
Carbon tax 40
Carboxyl concentration 24
Carbureted water gas 20
Carburetor 63
Carnotite 84
Carr & Tagore Coal Co. 6
Causticized lignite 98
Cell wool 61
Cement manufacturing 21
Cenosphere 37
Ceresin 67
Chadwick, James 85
Chain fission 88, 103
Chain reaction 88, 106
Char 9, 17
Characterization factor 56
Chemical fertilizer 59, 61, 77
Chloro-fluoro carbon (CFC) 24
Clay fly ash brick 36
Clean Air Act of USA 39
Coal bed methane (CBM) 25
Coal dust injection (CDI) 16, 27
Coal gas 6, 19
Coal mine methane (CMM) 27
Coal tar 12, 18, 74
Cobalt 60 93
Coke 5, 6
Coke breeze 29
Coke oven 9
Coking property 9
Combine cycle technique 79
Community Development Carbon Fund (CDCF) 40
Compressed natural gas (CNG) 73
Compression ratio 63, 64
Conductivity sorting 13, 14
Consumable welding rod 108
Corex 17
Cracking of petroleum 20, 55, 59, 62, 70, 77
Creosote oil 12,18
Critical size 92
Crucible swelling number 10
Curie, Irene 85
Curie, Marie 85
Curie, Pierre 85
Cyclone technology 13, 14
Cycloparaffins 52
Cymogene 57, 69

D

Daimler, Gottilab 53
Darbys, Abraham 5
Davy's lamp 19
Desulpho gypsum 31
Deuterium 96
Dielectric strength 57, 66
Diesel engine 64
Diesel, Rudolph 53
Dilatometric test 9, 10
Diolefines 55
Dios 17
Direct reduction technology 16, 17
Dispersion 104
Distillation 62
Domestic heating 20
Drake 53
Drake well 53
Dry beneficiation of coal 13, 14
Dry cleaning 67
Dry gas 75
Ductility 103
Dudley, Dud 5

E

East Indian railway 6
Edible fats 59, 61
Einstein, Albert 85, 88, 91, 96, 106
Eldorado 86
Electron emission 105, 107, 108
Electron tubes 105, 107
Electrostatic precipitator (ESP) 31
Emissivity 104, 107
Enhanced oil recovery (EOR) 70, 71
Ethanol 73, 77
Ethyl alcohol 73, 77
Ethylene 55
Explosives 59, 61
Extra heavy crude oil 70, 71

F

FAL-G technology 36, 40
Fast breeder reactor 90, 95, 105
Fast neutron 88, 94, 105
Fast reactor 94, 95
Ferghana 86
Fermi, Enrico 85
Field emission 105
Fine coal processing wastes (FCPW) 29
Fisher-Tropsch (FT) method 80
Fission 87, 88, 102, 105
Fission bomb 91, 92
Flash point 57, 72
Flat mirror collector system 42
Flat plate collector system 42
Fluidized bed combustion (FBC) 28
Fluorescent glass 93
Fly ash 31
Formed coke 26
Foulton, Robert 6
Foundry 24
Fractional crystallization 102
Fuel cell 19, 81
Fuel element 90
Fuel injection device 63,64
Fuel oil 69
Fuel ratio 8
Furnace bottom ash (FBA) 31
Furnace fuel oil 58, 69
Fusion bomb 96

G

Gamma phase of uranium 88
Gamma radiolytic process 14, 15

Gamma rays 87, 93
Gas Authority of India Ltd. (GAIL) 76
Gasohol 73
Gas oil 58
Gasoline 63
Gas-to-liquid (GTL) process 79, 80
Geological age determination 93
Geo-pressed water 45
Geo-thermal energy 44, 45
Gieseller plastometric test 9, 10
GKLT test 9, 10
Gobar gas 41
Gray-King test 9
Grease 69
Greencotton 68
Green house effect 30, 39
Green house gas (GHG) 30, 39, 40
Gross calorific value 8

H

Hahn, Otto 85
Half-life 87, 93
Hannay 53
Hazira-Bareilly-Jagdishpur (HBJ) Pipe line 76
Heavy crude oil 70
Heavy fuel oil 58
Heavy water 91
Helium 80
Helium Act Amendments 80
Herodotus 52
High speed diesel (HSD) 58, 64
High sulphur coal 26
Holland, Thomas 86
Hydroelectricity 44, 45
Hydrogen as fuel 72
Hydrogen bomb 96
Hydrothermal water 45

I

Imino 7
Incandescent light 93
Indian Rare Earths Ltd. (IREL) 100
Industrial revolution 39
Industrial waste heat 44, 51
In situ coal waste 24
Insulation bricks 36
Integrated combined cycle System 19
International Engery Agenty (IEA) 19, 31
Ionic island 52
Isomerization 62
Isotope 84

J

Jatropha oil 74
Joachimsthal 85, 86
Jolio, Federick Curie 85

K

Karanj oil 74
Kelly, Henry 46
Kerogen 70, 71
Kerosene 41, 58, 63
Klaporth, Martin Heinrich 85
Kyoto Protocol 40

L

Lamp filament 105, 107
Leco 40
Lenoir 53
Light diesel oil (LDO) 68
Liquefied natural gas (LNG) 78, 79
Liquefied petroleum gas (LPG) 57, 59, 65, 66, 78
Liquefied petroleum product (LPP) 78, 79
Liquid sodium 95

Locomotives 22
Lubricant 58
Lubricating oil 58, 66
Luminous painting of clock dial 93

M

Madras Oil Refinery 56
Magnetite in tailings from uranium concentrator 95
Malaria control 94
Maltha 52
Man-made fiber 59, 60, 61
Marsh gas 75
McKillop Stewart & Co. 53
Medlicott 53
Mesothorium 102, 103, 108
Met-coke 98
Methane hydrates 82
Methanogen bacteria 28
Methanol 73, 77
Methyl alcohol 73, 77
Meyers-Read process of desulphurization 26
Microwave processing of coal 14, 15
Mineral jelly 58, 67, 68
Mineral oil 52
Mineral pitch 52
Mini-OTEC 49
Model-making 67
Moderator 91
Mo-gas 57, 63
Mullite 37
Murdoch, William 6

N

Naptha 12, 18, 38, 43, 57, 61, 74
Naphthalene oil 12, 18, 55
Naphthenes 55
NASA 81
National Environmental Engineering Research Institute (NEERI) 80
National Physical Laboratory (NPL) 18
Natural bitumen 70, 71
Natural gas liquid (NGL) 79
Nebuchadnezzar 75
Needle coke 18
Net calorific value 8
Neucleon 103
Newcommen Thomas 6
Nitrogenous fertilizer 20, 61
Nitroglycerine 68
Non-consumable welding rod 108
Nuclear disarmament 95
Nuclear fuel 44, 89, 90
Nuclear fusion 96
Nuclear powered submarine 91
Nylon 61

O

Ocean thermal conversion (OTEC) 44, 48
Octane number 62, 63, 64, 65
Oil black 69
Oil from plastic 74
Oil India Ltd. 54
Oil shale 70, 71
Olefins 55
Oleoflotation 28
ONGC 54, 74
Optical glass 105, 108
Otto 53
Ozokerite 52, 67

P

Paraboloid mirror system 42
Paraffin 52, 54

Paraffin oil 58
Paraffin wax 58, 67
Pavement making 67
Peligot, Eugene Melchior 85
Perfumery 59, 61
Perlon 61
Pesticide 59, 61
Petroleum coke 58, 68, 71
Petroleum ethers 57
Petroleum gas 57
Petroleum pitch 58, 67
Phase transformation of uranium 88
Phenol 12, 18
Photoelectric emission 105
Photometry 69
Photometry sorting 13, 14
Photosynthesis 42
Photo-voltaic cells 46
Photo-voltaic programme 47
Pig Institute 98
Pitch 12, 18
Pitchblende 84, 85, 86, 89
Pit coal 5
Plastics 60
Plutonium 94, 95
Plutonium carbide 95
Poly-acrylic fiber 61
Polyester fiber 74, 77
Poly-ethelene terepthalate (PET) 74, 77
Polymerization 62
Polythelene 74, 77
Power generation 22
Prime coking coal 9
Printers' ink 69, 78
Printing ink 74, 78
Producer gas 19
Pseudomonas 29
Pulverized coal injection (PCI) 27
Pulverized fuel combustion 22
PVC 74, 77
Pyridine 12, 18

R

Radiator antifreeze 59, 61
Radioactivity 87
Radiotherapy 93, 105, 107
Radium 85, 93
Radium hill 86
Ramge 32
Resins 60
Rayon 61
Reactor 90
Reducing power 8
Reforming of petroleum 62
Refractive index 104
Rhodospirillum Rubrum 50
Road making 67
Rock oil 52
Romelt 17
Rotary Kiln 17
Rutherford 85

S

Salt gradient energy 44, 50
Sapozhnikov test 9, 10
Saturated hydrocarbon 54
Savery Thomas 6
Schomberg, C. W. 100
Scotch whisky 109
Shale-lime bricks 71
Shale oil 71
Shinkolobwe 86
Skinning 62
Sky lab 46
Slow neutron 88

Smelting reduction technology 17
Sodium lignite 98
Soft coke 17
Soil amendment 109, 110
Solar cell 46
Solar energy 44, 46
Solar heat 41
Solar hydrogen 83
Solar thermal electricity conversion (STEC) 46, 47
Solvent extraction method 102
Soyabean oil 73
Special Theory of Relativity 85
Spheroid grade iron 26
Sponge iron 16, 17
Spontaneous ignition 11
Sporotrichum Purverulesstum 28
Stamp-charged coke 26
Stamp-charging technology 26
Steam coal 22
Steel grade coking coal 16
Stephenson, George 6
Sterile Insect Technique (SIT) 94
Sterilization 93
Stoehr, Emil 86
Strassman, Fred 85
Stychnos Potatorium 94
Subcritical size 92
Sulphonyl concentration 24
Sulphur recovery from petroleum 70
Syncrude 80
Syngas 73, 80, 82
Synthesis gas 82
Synthetic detergent 59, 60
Synthetic fiber 60, 61, 74
Synthetic gypsum 31
Synthetic petroleum 23, 79, 82
Synthetic resin 59, 60, 74, 77
Synthetic rubber 59, 60, 74, 77
Synthetic zeolite 37
Swelling index 9, 10

T

Tagore, Dwarkanath 6
Tar sand 70, 71
Terylene 61
Tetra-ethyl lead (TEL) 64
Thermal neutron 88, 94, 105
Thermal reactor 90, 95, 105
Thermionic emission 105
Thermodynamic efficiency 61
Thermonuclear bomb 96
Thiobacillus Ferro-oxidant 26
Thor 100
Thorianite 100
Thorite 100
Tidal energy 44, 48
Toluene 12, 18
Topping 62
Torbenite 86
Tower of Babel 52
Trace elements in fly ash 32
Trinitro toluene (TNT) 61

U

Underground gasification of coal 25
Unit coal 7
Unsaturated hydrocarbon 55
Uraninite 84, 86, 95
Uranit 85
Uranium alloys 92
Uranium compounds 92, 93
Urgeiricia 86

V

Varnish 68, 74
Vaseline 67
Vertical shaft kiln 21

W

Washery grade coking coal 15
Water gas 19
Watt, James 6
Wave energy 44, 50
Wax 67, 68
Weatherability of coal 11
Welding 105, 108
Wet gas 75, 78
White coal 5
White, Major 53
White oil 58, 67
White spirit 58, 67
Wilcox 53
Wind energy 44, 47, 73
Wind energy conversion system (WECS) 48
World War-I 18, 100
World War-II 18, 25, 86
Wrocklaw Institute of Petroleum and Coal 99

X

Xylene 12, 18, 69, 74
Xylenol 18

Z

Zante 52